Graph Theoretic Algorithms for the Ground Based Strategic Deterrent Program

Prioritization and Scheduling

DON SNYDER, CHRISTIAN JOHNSON, PAROUSIA ROCKSTROH,
LANCE MENTHE, BART E. BENNETT

Prepared for the Department of the Air Force
Approved for public release; distribution unlimited

PROJECT AIR FORCE

For more information on this publication, visit **www.rand.org/t/RRA583-1**.

About RAND

The RAND Corporation is a research organization that develops solutions to public policy challenges to help make communities throughout the world safer and more secure, healthier and more prosperous. RAND is nonprofit, nonpartisan, and committed to the public interest. To learn more about RAND, visit www.rand.org.

Research Integrity

Our mission to help improve policy and decisionmaking through research and analysis is enabled through our core values of quality and objectivity and our unwavering commitment to the highest level of integrity and ethical behavior. To help ensure our research and analysis are rigorous, objective, and nonpartisan, we subject our research publications to a robust and exacting quality-assurance process; avoid both the appearance and reality of financial and other conflicts of interest through staff training, project screening, and a policy of mandatory disclosure; and pursue transparency in our research engagements through our commitment to the open publication of our research findings and recommendations, disclosure of the source of funding of published research, and policies to ensure intellectual independence. For more information, visit www.rand.org/about/principles.

RAND's publications do not necessarily reflect the opinions of its research clients and sponsors.

About This Report

The objective of this project was to develop quantitative methods to make the unified certification process for nuclear systems more rigorous and efficient. The emphasis was on the Ground Based Strategic Deterrent program because of its size and the availability of a rich data set from its models-based systems engineering digital environment. The approaches and algorithms presented in this report should serve as a template for future nuclear programs. This report should be of interest to the nuclear community in the U.S. Air Force, but the principles and algorithms are applicable to managing any complex process.

The research reported here was commissioned by the Commander of the Nuclear Weapons Center and Program Executive Officer for Strategic Systems and conducted within the Force Modernization and Employment Program of RAND Project AIR FORCE as part of a fiscal year 2020 project titled "Quantitative Measures of the Resiliency of Nuclear Weapon System Architectures."

RAND Project AIR FORCE

RAND Project AIR FORCE (PAF), a division of the RAND Corporation, is the Department of the Air Force's (DAF's) federally funded research and development center for studies and analyses, supporting both the United States Air Force and the United States Space Force. PAF provides the DAF with independent analyses of policy alternatives affecting the development, employment, combat readiness, and support of current and future air, space, and cyber forces. Research is conducted in four programs: Strategy and Doctrine; Force Modernization and Employment; Resource Management; and Workforce, Development, and Health. The research reported here was prepared under contract FA7014-16-D-1000.

Additional information about PAF is available on our website:
www.rand.org/paf/

This report documents work originally shared with the DAF on May 26, 2020. The draft report, issued on November 19, 2020, was reviewed by formal peer reviewers and DAF subject-matter experts.

Contents

Figures and Tables

Figures

Tables

Summary

Issue

As programs grow larger and the interdependencies of the program activities get complex, tools are needed to introduce more rigor into program management to reduce schedule risk. In this report, we present the underlying rationale behind algorithms already delivered to the Ground Based Strategic Deterrent (GBSD) program office that were designed to reduce the likelihood of rework in program execution, provide better insight into schedule risk, and provide insights into how to restructure task dependencies to manage schedule risk.

The algorithms help project managers consider "what if" scenarios and visualize potential pitfalls as a project progresses. Project managers can also use these algorithms to perform sensitivity analyses on individual activities to understand which activities have the greatest potential to derail a project. Although these algorithms were developed for the GBSD program office, specifically for the Toolbox for the Unified Certification Strategy Dashboard, the methods are general and applicable to any program.

Approach

This report builds on the project management methods of Program Evaluation Review Technique (PERT) and Critical Path Method in two novel ways. The first is to show how the critical path for execution depends on the overall structure of how the tasks are scheduled. The mathematical representation is called a *graph*, and the structure is called the *graph topology*. Monte Carlo simulations have been used for decades to find the critical path for schedule completion. Unfortunately, simulation-based methods become impractical for large graphs because of computation times. The second novelty is to present semi-analytical methods that are orders of magnitude faster than Monte Carlo simulations and can be used efficiently on very large graphs. We also explore how the reallocation of resources can change the risk of meeting expected schedules.

Findings and Conclusions

Novel Topological Insights

The most probable critical path and degree of slack for project completion depend on graph topology. When managers have the opportunity to change the structure of activities, they can better manage where schedule risk lies in program execution. We examined three classes of graph topology: a simple topology defined by the number of parallel chains in the graph, random graphs that display Erdős-Rényi topology, and scale-free graphs.

The simple chain topology is elementary but contains many structural elements common to many PERT graphs. It is intuitive and contains the basic structural forms that a manager can readily adjust. The random and scale-free graph cases are common topologies in naturally occurring graphs. Many large graphs present one of these two topologies; therefore, they are useful baselines for analysis. We find the following:

- For the simple chain topology, although parallelizing tasks can theoretically reduce the overall time, it does so less than might be expected, because the likelihood of having at least one bad chain increases with the number of chains. There is a trade-off: Parallelizing development reduces the variance more than expected but increases the end-to-end time more than expected. Thus, although working tasks in parallel remains desirable, the project manager's expectations for time savings should be tempered.
- For Erdős-Rényi topology (random graphs), our analysis shows the same trade-off between variance and average completion time as seen in the simple chain topology, though the effect is more subtle. In this case, the presence of interlinking paths makes it less likely that a single chain will dominate, because the chains are not independent.
- For the scale-free graph topology, our analysis shows that graphs with a single, dominant longest path are much more likely to complete on time than graphs with several parallel paths that may each be the critical path. This result means that, for scale-free graphs, the project manager can reduce the overall completion time with only a marginal increase in the corresponding risk by placing important tasks in series with one another.

Novel Numerical Methods

Although Monte Carlo techniques are useful for finding the critical path in a small PERT graph, as graph size increases, these methods are too computationally slow to be of practical use. To overcome this limitation, we present two numerical approaches that are many orders of magnitude faster:

- The fastest—and preferred—method transforms the problem into a new space with Chebyshev polynomials as a basis. The relevant calculations are performed in that space; then, the results are inverted back to the original space.
- The other method uses trapezoidal integration but is preferred only when software packages limit the ability to manipulate Chebyshev polynomials.

Resource Constraints and Future Analysis

We found three main resource constraints and opportunities for future analysis:

- We describe how slack in a task network can be manipulated to reduce project completion time; to reduce the overall need for resources; or to respond to other needs, such as start-up constraints, changes in workforce availability or capacity, or cash flows.

- Current methods rely on either expected-value algorithms, which may not sufficiently capture uncertainty, or Monte Carlo simulations, which require excessive amounts of runtime for large, complex projects.
- Future work in this area could extend the novel numerical methods described in this report to closely approximate solutions that could prove invaluable to the program manager.

The algorithms that have been delivered to the GBSD program office and this explanation of those approaches should aid in the management of complex programs, especially those following digital engineering practices.

Acknowledgments

We thank Maj Gen Shaun Morris and Brig Gen Anthony Genatempo for sponsoring the work. Col Jason Bartolomei provided day-to-day guidance and intellectual inputs into the algorithms. (Ranks are those at the time of this research.) We also deeply appreciate the collegial interactions with the entire Ground Based Strategic Deterrent mission defense team, whose members are too numerous to cite individually. Discussions and feedback from the Air Force Nuclear Weapons Center and the Air Force Safety Center are greatly appreciated and enhanced our understanding of the relevant processes. Our colleagues at the MITRE Corporation were very helpful, providing insights into unified certification challenges and feedback on our work. We especially thank Laura Antul and Tim Watt for many discussions and their insights.

At the RAND Corporation, we thank Gary Briggs for review of and comments on the computer code that instantiates this analysis. We thank Paul Dreyer for ideas and insights early in the project and Sean Colbert-Kelly and Jim Quinlivan for formal reviews that strengthened the work.

That we received help and insights from those acknowledged above should not be taken to imply that they concur with the views expressed in this report. We alone are responsible for the content, including any errors or oversights.

Abbreviations

AoA	activities on arcs
AoN	activities on nodes
CDF	cumulative distribution function
CPM	Critical Path Method
DAG	directed acyclic graph
GBSD	Ground Based Strategic Deterrent
iid	independent and identically distributed
KHF	Kummer confluent hypergeometric function
LCT	latest cumulative time
MBSE	Model-Based Systems Engineering
MOS	maximum order statistic
PDF	probability density function
PERT	Program Evaluation Review Technique
UCS	Unified Certification Strategy

1. Introduction

Background

During the execution of programs, managers must orchestrate the planning and performance of upwards of tens of thousands of interdependent tasks.[1] Many of these tasks have interdependencies such that failure to meet scheduling expectations in one task can affect the execution of many other tasks. The larger the program and the more complicated the interdependencies, the more difficult it is to detect potential cascading schedule slips in time to mitigate them. Failure to actively manage cascading schedule slips can jeopardize the timely execution of the entire program.

The Ground Based Strategic Deterrent (GBSD) currently under development is a good example of a large, complex program with numerous interdependent tasks. GBSD is a complete replacement for the aging LGM-30 Minuteman III intercontinental ballistic missile system. This program will design and produce more than 600 missiles, refurbish 450 operationally configured launch facilities (silos), design and produce at least 24 launch control centers, create the requisite weapon system command and control, and produce numerous support equipment and facilities for test and operations support. All of these activities will dovetail with the decommissioning of the Minuteman III while meeting continuous national strategic nuclear deterrence objectives.

As a nuclear program, GBSD will require rigorous cybersecurity and cyber resiliency risk assessments,[2] nuclear certification,[3] and nuclear safety certification. These three sets of activities overlap enough that the GBSD program office developed a Unified Certification Strategy (UCS) to integrate the overlapping tasks. Given the tight schedule by which GBSD is bound and the numerous organizations participating, it is necessary that these three sets of activities proceed in an efficient manner. To assist this management, a UCS Toolkit is under construction to support a Dashboard for all of these interrelated activities.

Increasing digitization presents an opportunity for better management of these numerous, interdependent activities. GBSD is a pathfinder program for digital engineering.[4] Digital

---

[1] In the management literature, the terms *task* and *activity* are used interchangeably. We will not make any distinction between the two terms in this report.

[2] Department of Defense Instruction 8510.01, *Risk Management Framework (RMF) for DoD Information Technology (IT)*, Washington, D.C.: U.S. Department of Defense, March 12, 2014, incorporating change 2, July 28, 2017.

[3] Air Force Instruction 63-125, *Nuclear Certification Program*, Washington, D.C.: Headquarters, Department of the Air Force, January 16, 2020.

[4] Office of the Deputy Assistant Secretary of Defense for Systems Engineering, *Department of Defense Digital Engineering Strategy*, Washington, D.C.: U.S. Department of Defense, June 2018; Phil Zimmerman, Tracee Gilbert, and Frank Salvatore, "Digital Engineering Transformation Across the Department of Defense," *Journal of Defense*

engineering holds the authoritative data for a program in a common digital format, enabling the use of common tools and models to support a range of activities across the life cycle of a program. One element of digital engineering is to hold acquisition artifacts in digital form in this environment rather than as stand-alone documents and spreadsheets. Digital engineering in the GBSD program facilitates tool development to better manage the execution of numerous activities that have interdependencies.

We developed and delivered modeling tools for the GBSD program office based on graph theory to help manage the UCS process. We worked with the MITRE Corporation to implement the computer code as part of the UCS Toolkit in the GBSD UCS Dashboard.[5] This report presents the underlying logic and mathematics behind these algorithms. While the algorithms and the corresponding computer code are particularly of interest to the GBSD program office, the generality of these methods and the standardization of digital engineering architectures allow them to be of use to other programs.

We describe a set of algorithms based on graph theory (of which we explain the basic concepts in the last section of this chapter) that is designed to help prioritize UCS tasks, estimate the probabilities of schedule slippage, and show how to make these decisions under resource constraints. These tools help project managers consider "what if" scenarios and visualize potential pitfalls as the project progresses. They can also help program managers perform sensitivity analyses on individual activities to gain insight into which have the greatest probability to derail the project.

Despite being developed for UCS tasks, the methods are general and are useful for any large set of tasks with interdependencies. We also describe the advanced mathematical methods necessary to implement these algorithms in a practical way for very large graphs. The purpose of this report is to provide decisionmakers with a high-level description of the methodology and metrics that we have developed to improve their ability to visualize and manage complex projects, including GBSD. This report is written to a more technical audience that is interested in the algorithmic details and mathematics behind these methods.

Structure of This Report

This report documents several algorithms for analyzing task-dependency graphs to inform task prioritization, scheduling, and related factors. We present them in this order because the algorithms build on one another. Chapter 2 describes algorithms for prioritizing tasks, along with some associated error checking. Chapter 3 describes algorithms for estimating schedule risk. Chapter 4 describes algorithms for adding resource constraints to the scheduling and

---

Modeling and Simulation: Applications, Methodology, Technology, Vol. 16, No. 4, 2019; and Will Roper, *There Is No Spoon: The New Digital Acquisition Reality*, September 18, 2020.

[5] The algorithms are designed to be implemented using Python in MITRE'S CAMEO plug-in and React application, which leverages Pointillist and Dagger for visualization.

prioritization problems. Chapter 5 explores an important parameter that emerges from this analysis, the criticality index, and addresses computational issues for conducting these calculations on large graphs. Finally, Chapter 6 discusses broader implications and next steps.

Basic Concepts

Although in common usage a *graph* can refer to any kind of plot, we use the term in the mathematical sense in this report. We define a *graph* as a collection of objects called *nodes* (or vertices) joined by links (or edges).[6] In this report, all graphs are *directed*, meaning that each link points from one node to another node (and none are bidirectional), and all graphs are *simple*, meaning that any pair of nodes may be joined by at most one link. A typical representation of a directed simple graph is shown in Figure 1.1.[7]

Figure 1.1. A Directed Simple Graph

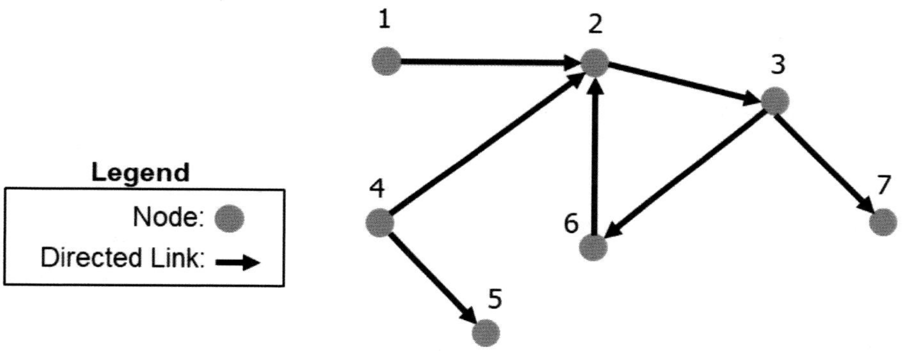

As shown in Figure 1.1, we may assign unique identifiers to each node (usually numbers or letters) and represent the links as ordered pairs of these nodes. Note that the positions of the nodes and the lengths of the links in this diagram are irrelevant: All that matters is how they are connected. Thus, a mathematically complete description of the graph depicted in Figure 1.1 would be the following set of ordered pairs of nodes: {{1, 2}, {2, 3}, {4, 2}, {4, 5}, {6, 2}, {3, 6}, {3, 7}}.

The power of graph theory derives from this simple representation of complex relationships. Graphs can be used to represent almost any kind of structured relationship, from electric circuits to financial transactions to the World Wide Web.[8] Therefore, many different graphs may be

---

[6] Different fields of mathematics use different terms. We use *nodes* and *links*, as is done in network analysis. For more on graph theory, there are many choices. For example, a standard text is Reinhard Diestel, *Graph Theory*, 5th ed., Berlin, Germany: Springer, 2017.

[7] Directed simple graphs are sometimes referred to by the abbreviation *digraph*.

[8] David Easley and Jon Kleinberg, *Networks, Crowds, and Markets: Reasoning About a Highly Connected World*, New York: Cambridge University Press, 2010.

abstracted from GBSD UCS-related data, representing such relationships as administrative control, policy linkages, power couplings, shared system components, and network connections. Once these relationships are represented as a graph, a powerful set of mathematical tools is available to analyze them. With well-devised graph-theoretic algorithms, we may then analyze these graphs to obtain insights about these relationships to assist UCS management. The key is to apply the right algorithms to the right graphs to obtain the desired insights. Figure 1.2 shows this analysis cycle.

Figure 1.2. Using Graph Theory to Inform UCS Planning

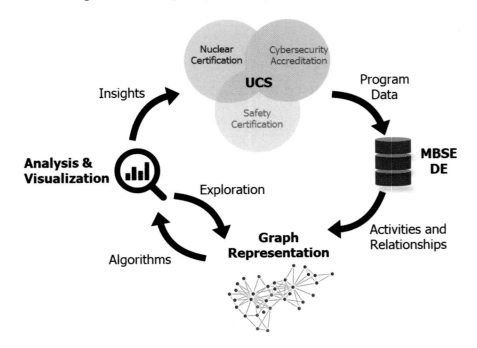

NOTE: DE = digital engineering; MBSE = Model-Based Systems Engineering.

In this report, we use graphs to represent certification task dependencies. The nodes in these graphs symbolize generic nuclear certification activities, such as tests, approvals, consultations, meetings, and milestones. The links represent logical dependencies between these activities. To extract these graphs from the underlying data, we make four key assumptions in the process.

First, we assume that each task has a clearly defined beginning and end. Second, we assume that each dependency applies to the end of one task and the beginning of another. For example, if a component requires testing, but it cannot be tested until a prototype has been built, we may extract from this description two task nodes—"building the prototype" and "testing the component"—and one dependency link between them representing a logical succession: The build must be complete before the test begins. As per the usual definition, a *predecessor* node is one that comes before; a *successor* node is one that comes after.

4

Of course, we do not expect that all task dependency information will be so simple. Some data cleaning or conditioning will likely be needed. Some tasks may be defined too broadly for our purposes, incorporating many parts, and we may need to subdivide them to make the dependency relationships cleaner. Likewise, to make the dependency relationships cleaner, we may need to split those as well. Nevertheless, this graph representation approach is quite general and broadly applicable to many processes.

The third assumption is that all tasks belong to some development path—a sequence of nodes that originates from a common start point and concludes at a common end point. In other words, the certification process is unified. Mathematically, this ensures that the graph is connected, which allows us to make use of several mathematical theorems.[9] Fortunately, for the system we are considering, we can impose this condition without loss of generality. If needed, we may add a new start node that precedes all other activities and a new end node to the graph that represents completion of all activities. With these, we can then add links, such that the new start node joins to any tasks without predecessors, and any tasks without successors join to the new end node. This procedure works in general, and the result is unique. The procedure is illustrated in Figure 1.3.

Figure 1.3. Creating a Connected, Directed Graph

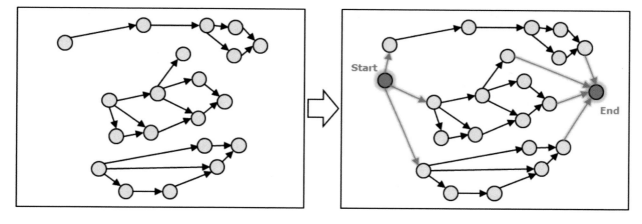

NOTE: The left panel shows a disconnected, directed graph; the right shows the addition of common start and end nodes.

Finally, we assume that the final graph has no loops, or that we may collapse any extant loops into single tasks.[10] Loops are problematic for many graph theory algorithms. They also may indicate inconsistencies in policy or process design. Of course, the presence of circular dependencies does not necessarily indicate inconsistencies—cycles are part of many Air Force

---

[9] A *connected graph* is one in which every node can be reached via a series of hops from every other node, ignoring the directionality of the arrows.

[10] We also assume that no node links back to itself—that the start and end nodes of a link are always distinct.

processes—but for an acquisition process constrained by time and resources, iterative processes cannot continue indefinitely. If the number of cycles is known, we may represent each as a separate task, but if the number of repetitions is not fixed, we must represent the loop as one (complex) task. The formal process of removing loops from a graph in this manner is known as *condensation*, and the resulting graph is called *acyclic*.[11]

The result of this four-part abstraction process is that we may represent the task dependencies in the UCS as a *simple, connected, directed acyclic graph* (DAG).[12] This is an important class of graphs that has been well studied. Figure 1.4 shows a notional DAG in a common format that we use and refer to throughout this report. Note that the identifiers for each node—in this case, letters—are inscribed in the nodes.

Figure 1.4. Notional Task Dependency Graph

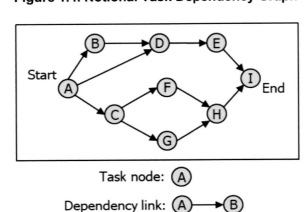

Task node: (A)

Dependency link: (A)——▶(B)

---

[11] Robert E. Tarjan, "Edge-Disjoint Spanning Trees and Depth-First Search," *Acta Informatica*, Vol. 6, No. 2, 1976.

[12] Although all DAGs are simple, they need not be connected—but all the ones in this report will be.

2. Prioritization

There are many strategies for prioritizing certification tasks for completion—that is, to indicate the order in which they should be worked. One rationale for prioritizing tasks is to reduce the risk of rework.

In the context of an acquisition program, *rework* means having to redo tasks that have already been successfully completed. This can be costly in terms of time and resources. The rework that we consider here does not include the additional effort required to rectify the kinds of normal mistakes anticipated during operations, such as occasionally having to recast a part because of a temperature-control problem in the foundry or occasionally needing to repeat a cleanliness test because of accidental contamination of the sample. We account for anticipated variances in the next chapter when we talk about the probability that a given task may take more time than expected. *Rework* instead refers to the need to redo entire tasks because of unanticipated changes in task requirements, such as those that might be needed to address a serious design flaw that comes to light late in the process.

Although, by the very definition of *rework* as the result of an unanticipated event, we can never avoid all risk of rework, we can avoid *some* risk of it by judiciously ordering tasks to reduce the risk of rework cascades. This practice is common. For example, when repairing a desktop computer, one does not screw the lid back on until one has verified that the system is working correctly to avoid having to remove and replace the cover twice. Likewise, one does not mop the kitchen floor until after one has cleaned the counters. Some tasks have the potential to affect more tasks down the line than others; therefore, they should, if possible, be completed first.

In this chapter, we generalize this principle using the graph theoretic concepts of *reach* and *topological order*. The algorithms we describe to compute these measures pave the way for the more general discussion of scheduling algorithms in the next chapter.

Reach

Intuitively, the reach of a task is the number of tasks that depend on it. The reach of a task, therefore, represents the number of other tasks that would have to be redone if the task must be redone. For a DAG, the *reach* of a node (say, node A) is defined as the number of nodes that lie on at least one path that begins at that node (in this case, node A). Reach is easy to visualize using the task dependency graph introduced in the previous chapter. Figure 2.1 illustrates the concept. In this graph, the highlighted node (E) has a reach of five because five nodes (F, I, J, O, and Z) can be reached from that task following allowed paths (shown in red). For a DAG, the reach of every node is computable and uniquely defined.

Figure 2.1. Example of Reach

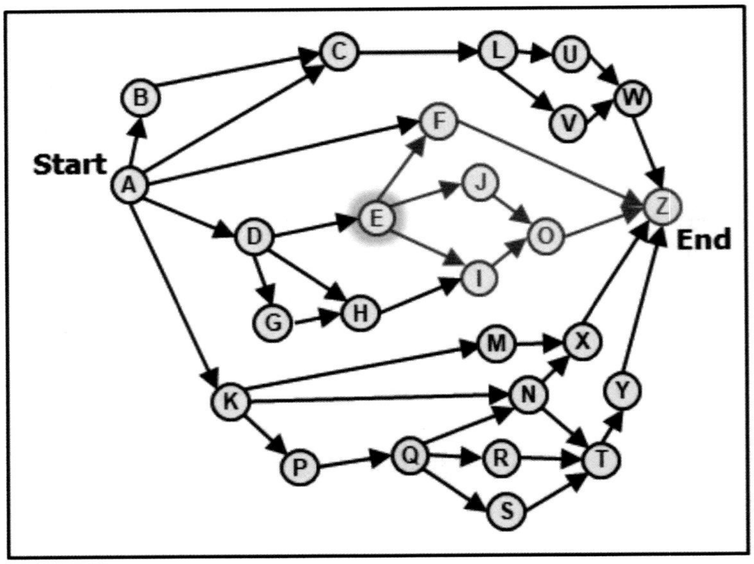

NOTE: In this graph, node E (highlighted and red) has a reach of five (the other red nodes).

The algorithm is straightforward, but it is not quite as simple as it might seem because one must ensure that each successor node is counted only once, even when the node may be reached via multiple paths. Reach counts each node equally, although some nodes may be more important than others.[13]

Prioritization in terms of reach would mean performing the tasks with highest reach first. The ones with the highest reach are the linchpins of the process. Ordering the tasks in this manner minimizes the risk of unnecessary rework if unforeseen circumstances arise. However, many tasks may have the same reach, creating ambiguity. The related concept of topological order, which we introduce next, suffers from similar ambiguities, but it is a more general foundation for the methods we introduce later.

Topological Ordering

By carefully constructing a task dependency graph as a DAG following the assumptions and process described in Chapter 1—by ensuring that all task nodes and dependency links are clearly defined without loops or other degeneracies—we can leverage important theorems about DAGs. One key theorem concerns the existence of topological ordering.[14]

A topological ordering is a valid and complete sequence of nodes. It is valid in the sense that it obeys all of the dependency rules represented by the links—i.e., when following this sequence,

---

[13] Although we could allow varying weights, the need for this is obviated by the more advanced algorithms in Chapter 3, where we use estimated completion time as the effective weighting scheme.

[14] Topological ordering is also called a *topological sort*.

one would never traverse any arrows in reverse. It is complete in the sense that it includes every node. Mathematically, a topological ordering of a graph is defined as a sequence of all nodes such that for every link {j, k}, node j comes before node k. Every DAG is guaranteed to have at least one topological ordering.[15] Figure 2.2 illustrates how a DAG might be unfolded to reveal its topological ordering.

Figure 2.2. Example of a Topological Ordering

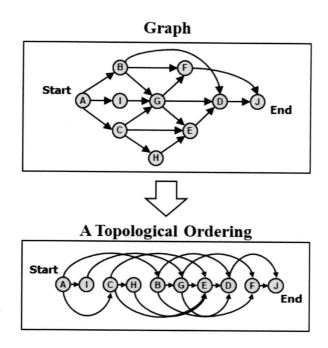

Note, however, that we say that every DAG has *at least one* topological ordering. Nodes that lie far apart on different branches may be sequenced differently in different topological orderings. This is natural and desirable, as it means that these tasks may be completed safely in any order—or, better yet, that they may be completed at the same time in parallel, resources permitting. As we will see in Chapter 4, where we consider resource constraints, a graph that admits multiple topological orders offers greater flexibility while still minimizing the risk of inducing a rework cascade. The number of possible topological orders of tasks is a measure of program execution flexibility.

Some subsets of nodes, however, always appear in the same sequence in all topological orderings. We call these nodes *strictly ordered*. This sequence of tasks often includes a series of benchmarks or milestones. Subsets of nodes that are not strictly ordered represent relatively

---

[15] Dieter Jungnickel, *Graphs, Networks, and Algorithms*, 4th ed., Berlin: Springer-Verlag, 2013, p. 50.

independent segments of development paths. When we consider scheduling issues in the next chapter, we observe that subsets that are not strictly ordered may afford slack in the process.

The concept of topological ordering is well understood, and algorithms that compute topological orderings are guaranteed to run in linear time, meaning that the execution time scales in direct proportion to the number of nodes and links.[16]

Error Checking

Finally, we note that the algorithms for computing reach and topological ordering have a useful side benefit for error checking. The graphs that represent GBSD data may include many thousands of certification tasks, and it is to be expected that there may be some errors in the data set. When a task dependency graph is constructed and reach and topological ordering are computed, certain types of incomplete or inconsistent relationships within the data become evident. Figure 2.3 illustrates some of the menagerie of defects that may be identified in this manner.

Figure 2.3. Types of Graph Defects

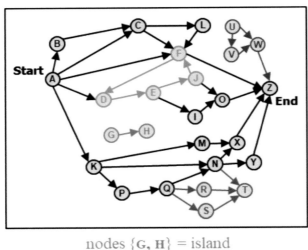

nodes {G, H} = island
nodes {U, V, W} = unrooted
nodes {R, S, T} = unfinished
nodes {D, E, J, F} = loop

The reach algorithm will naturally identify any *islands*, or isolated sets of nodes. Any node (or branch) that has no predecessors is *unrooted*; any node (or branch) that has no successors—

---

[16] Arthur B. Kahn, "Topological Sorting of Large Networks," *Communications of the Association for Computing Machinery*, Vol. 5, No. 11, 1962.

that has zero reach—is *unfinished.* Similarly, topological ordering identifies dependency loops. If any circular dependencies remain, the graph has no topological ordering.

As a practical matter, therefore, we anticipate that these algorithms will be used iteratively to adjust and create the initial task dependency graph itself. There is no simpler way to identify the very defects that must be removed to make this analysis possible. This approach can also reveal conceptual issues with the definitions of tasks and overlapping requirements that create a snarl instead of a clean DAG. That is one reason why we present these concepts in this separate chapter, even though the scheduling algorithms that we present next—and that form the heart of this report—necessarily already incorporate them.

3. Scheduling

Projects that involve a large number of interrelated tasks are sufficiently complex that they require advanced mathematical methods to measure progress and compute completion time metrics valued by decisionmakers. These metrics include how long it should take to complete the entire project, what fraction of the project is completed, and the timing implications of delays or uncertainties in various tasks. In this chapter, we begin with a brief overview of the genesis of these methods. We then apply these methods to the task dependency graphs previously described and illustrate some of the challenges involved with these methods.

Origins of Project Management Algorithms

Large-scale projects, from the building of the Egyptian pyramids to the construction of the Burj Khalifa in Dubai, have required careful management of the many interwoven activities necessary to perform them. Formal algorithms for laying out and evaluating some project management problems have their origins only 70 years ago, during the project management of the U.S. Navy's Polaris program.[17] Methods developed for Polaris were later developed into a body of work known as the *Program Evaluation Review Technique* (PERT).[18]

The formalism began with an abstraction of the many project tasks or activities and their precedence relationships into a graph with the activities represented by the nodes in the graph and the precedence relationships as the links.[19] Each task in PERT is also specified with estimations of the most likely, optimistic, and pessimistic times to finish the task. These estimates are combined statistically to create an approximate average time—an expected value— to complete the task.

Once this information is collected, PERT can be applied to provide decisionmakers with significant information during both the planning and execution phases of a project. First, the algorithm calculates the minimum length of time needed to complete the project. This includes opportunities to shorten the completion time by processing in parallel those activities that are not dependent on each other. Second, it identifies those nodes that, if delayed, will cause the whole project to be delayed. These nodes are said to be on the *critical path* for the project. Third, PERT

---

[17] The Polaris program was part of the Fleet Ballistic Missile Program. For a thorough review, see Harvey M. Sapolsky, *The Polaris System Development: Bureaucratic and Programmatic Success in Government*, Cambridge, Mass.: Harvard University Press, 1972.

[18] Mats Engwall, "PERT: Polaris and the Realities of Project Execution," *International Journal of Managing Projects in Business*, Vol. 5, No. 4, 2012.

[19] When PERT was first developed, tasks were represented by links and nodes denoted starting or ending points for the tasks. More recently, the typical abstraction places tasks on the nodes. This representation simplifies the network and can be constructed as a functionally equivalent network. The equivalence is shown in Appendix A.

calculates the length of time that tasks not on the critical path can be delayed without negatively affecting the project minimum completion time. These tasks are said to have *slack* and may be useful to decisionmakers in considering options for reallocating resources to reduce the time required to finish some task on the critical path and thus reduce overall project completion time.[20] Finally, a target completion date can be given for the project, and PERT can determine those tasks that must be reduced in time to achieve the overall project deadline. For additional details on the expected-value PERT algorithm, see Appendix B.

PERT has been lauded as the project management tool that was primarily responsible for the completion of the Polaris program earlier than expected.[21] Coincidentally, around the same time that PERT was being developed by the Polaris program, a similar methodology was developed by mathematicians at DuPont.[22] Their Critical Path Method (CPM) had as its objective the reduction in costs of plant shutdowns and restarts by better scheduling tasks. CPM also used a network as an abstraction of the tasks that needed to be completed in a large project. It included in its method an ability to look at cost trade-offs with project end time.

After the success of Polaris, the acceptance and popularity of PERT as a significant tool for the project manager grew. Over time, because their underlying abstractions were identical, PERT and CPM were frequently merged to take advantage of the best of both techniques. Today, a variety of software vendors, such as Microsoft, Workfront, Monday, and Jira, produce software products for project management based on the PERT/CPM algorithms.

Our analysis grows out of and builds upon the PERT/CPM literature, but we do not limit ourselves to the methods used by the original authors.

Estimating Times

The task dependency graphs discussed in Chapters 1 and 2 are completely described as sets of nodes and links. We now add timing parameters to the nodes. Mathematically, this is a "decoration" of the underlying graph that does not alter its structure. Following the PERT methodology, we apply three weights to each task node: a *most likely* completion time, an *optimistic* or minimum completion time, and a *pessimistic* or maximum completion time. With

---

[20] In a very large network, it may be difficult to determine what resources should be moved from which slack task to which tasks on the critical path or whose expected completion time will delay the overall time to finish the project. Additional methods beyond the PERT algorithm can assist decisionmakers in selection of these preferred tasks.

[21] Although PERT was primarily developed during the Polaris program, some believe that the success of Polaris was more attributable to the overall management strategy of the program or the plentiful availability of resources. See Sapolsky, 1972; and Mats Engwall, *PERT, Polaris, and the Realities of Project Execution*, Stockholm, Sweden: KTH Royal Institute of Technology, 2015. We also note that factors external to a project can play as significant a role or more-significant roles in determining project success. See Mats Engwall, "No Project Is an Island: Linking Projects to History and Context," *Research Policy*, Vol. 32, No. 5, May 2003.

[22] James E. Kelley, Jr., and Morgan R. Walker, "The Origins of CPM: A Personal History," *PM Network*, Vol. 3, No. 2, February 1989.

these parameters, we construct a probability density function (PDF) for the completion time of each task. Following the PERT method, the particular PDF we use is a type of beta distribution called the *four-parameter beta distribution*. The mathematical expression for the distribution is

$$P(t; \alpha, \beta, o, p) = \begin{cases} o \leq t \leq p: & \dfrac{(t-o)^{\alpha-1}(p-t)^{\beta-1}}{B(\alpha, \beta)(o-p)^{\alpha+\beta-1}}, \\ t < o \text{ or } p < t: & 0 \end{cases}$$

where $\alpha = 1 + 4\frac{m-o}{p-o}$, $\beta = 1 + 4\frac{p-m}{p-o}$, o = the optimistic time, m = the most likely time, p = the pessimistic time, and $B(\alpha,\beta)$ is the beta function, also known as the *Euler integral of the first kind*. Figure 3.1 shows the general form of this function.

Figure 3.1. Beta Distribution

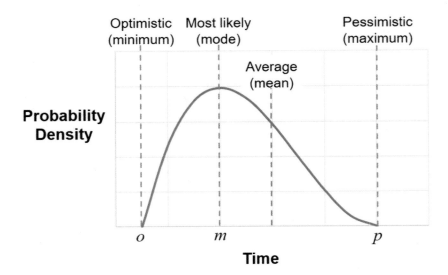

As is evident in Figure 3.1, the function (shown in blue) is bounded and unimodal. Because it is bounded, it is exactly zero for times less than o and greater than p.[23] Because it is unimodal, there is only one most likely completion time. The most likely time could be the average of o and p. In that case, the density function is symmetric about the mean. This special case, the symmetric beta distribution, requires only the optimistic and pessimistic times to be fully

---

[23] This is an advantage over other possible choices, such as the normal distribution, which would allow small but finite probabilities of negative and near-infinite duration.

specified.[24] Figure 3.1, however, shows an example of a nonsymmetric case. This example is a right-skewed beta distribution and is representative of a case in which the most likely time is closer to the minimum than the maximum. The mostly likely estimate could just as well be closer to the maximum than the minimum (left skewed). The PDF is also properly normalized.[25]

It should be noted that any representation of expected task durations is only as good as the input data. Estimates for optimistic and pessimistic times are salient examples. If there are systematic biases in the estimates of these parameters, the inferences from the analysis will also be biased. Some workers might underestimate the time to complete a task, not wanting to signal that they need a lot of resources, or perhaps out of fear of the attention of being flagged as the rate-limiting step. Others might overestimate the time needed in a ploy to advocate for additional resources. Or workers might just not have the experience and knowledge to accurately estimate these times. Although the implications of biased parameter estimates are important, we do not further analyze them.

There is no *a priori* reason for estimating the PDFs with beta distributions. The beta distribution has been used by previous analysts because it is a simple mathematical function that satisfies the conditions of being continuous and unimodal, intersects the abscissa at the optimistic and pessimistic values, and is sufficiently well behaved mathematically to be amenable to a range of manipulations. Given the uncertainties in estimating the parameters of the distributions and the cancellations of differences when assessed throughout the graph, there is no compelling reason to use other distributions.[26] Although the methods we describe hereafter would work just as well with any such probability distribution, we use this particular function for these reasons.

The completion time for each path in a graph can be calculated by computing the convolution of the respective PDFs for the beta distributions at each node along the path. In the case for which all nodes have the same distribution, and assuming that the distributions are independent, the distribution of completion times for each path converges to a normal distribution as the path length increases. This result can be used to compare and analyze paths through the graph. For a proof, see Appendix C.

Of course, the graphs that can be constructed are limited by the data available. The scheduling analysis described in this report requires completion time estimates for each task. In the case in which only the most likely completion time has been estimated for a task, we may add a probability envelope (e.g., ±20 percent variability) and perform parametric analysis with

---

[24] This is a special case of the four-parameter beta distribution in which $m = \frac{o+p}{2}$. Setting the scale parameters to $\alpha = \beta = 3$, a simplified, symmetric form of this distribution is $P(t; o, p) = \begin{cases} o \leq t \leq p: & \frac{30}{(b-a)^5}(t-o)^2(p-t)^2 \\ t < o \text{ or } p < t: & 0 \end{cases}$.

[25] Mathematically, the integral of the PDF over all time equals 1.

[26] K. R. MacCrimmon and C. A. Ryavec, *An Analytical Study of the PERT Assumptions*, Santa Monica, Calif.: RAND Corporation, RM-3408-PR, 1962.

that value to explore the effects. However, actual estimates including best- and worst-case times are more desirable.

We implement this algorithm in two ways. The first method uses a Monte Carlo method based on previous RAND Corporation work.[27] This approach is more intuitive but is computationally intensive and requires many replications to achieve statistical significance and a criterion for convergence. The second method introduces a novel analytic approach, which requires more-elegant mathematical machinery but needs to be run only once. As we show later, the analytic approach is preferred for large graphs. For pedagogical purposes, however, we follow the Monte Carlo method in the remainder of this chapter and the next. We turn to the analytic approach in Chapter 5.

The Critical Path

Once we have constructed a task dependency graph with probability distributions for the completion times of each task, we can simulate an instance of the graph to see what the overall schedule might look like. To run a simulation of the graph means to conduct random draws for the completion times of each task. In doing so, we implicitly assume that the random draws for each task are independent and that each task starts as early as possible—i.e., as soon as all predecessor tasks have been completed—so some tasks may run in parallel. We will discuss how and why we might relax these assumptions in the subsequent chapters. We may run the graph as many times as desired to compute various statistics, such as estimates of the average end-to-end completion time of the process and the standard deviation of that time.

One important concept that emerges from this type of analysis is the notion of the critical path. The critical path is the proverbial long pole in the tent: It is the end-to-end sequence of tasks that takes the longest amount of time. Delays along the critical path will therefore delay the project as a whole, while delays along other paths are unlikely to derail the schedule.[28] Identifying the critical path, therefore, can help the project manager decide how to allocate resources to prevent costly scheduling overruns. It also provides a useful benchmark, as it represents the shortest time in which one could reasonably expect to complete all tasks, if resources were not limited. For a DAG, the algorithm to identify the critical path for an instance of the graph runs in linear time using topological ordering.[29]

Of course, depending on the random draws, the longest path for any particular instance of the graph may change. Furthermore, as the certification process progresses, the remaining critical path may change (if, for instance, a particularly slow node unexpectedly finishes ahead of

---

[27] Richard M. Van Slyke, "Monte Carlo Methods and the PERT Problem," *Operations Research*, Vol. 11, No. 5, September–October 1963.

[28] In the rare event that multiple paths share the same longest time, they are all equally critical paths.

[29] The longest path for a DAG can be found in $O(n)$ time using a topological ordering. See Joseph J. Moder and Cecil R. Phillips, *Project Management with CPM and PERT*, New York: Van Nostrand-Reinhold, 1970.

schedule). Figure 3.2 illustrates how we can use these methods to analyze the shifting schedule and critical path for simple graphs.

Figure 3.2. Example of Shifting Schedules and the Critical Path

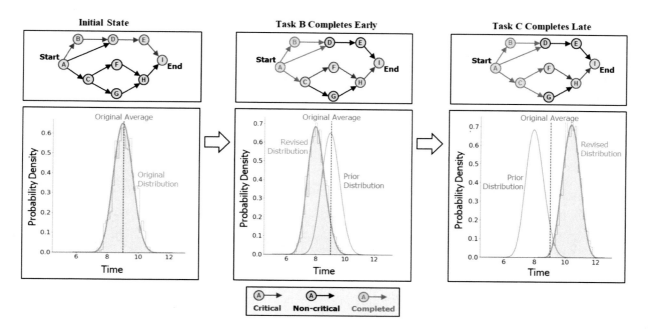

NOTE: The most likely, maximum, and minimum completion times for each node are not shown.

This figure shows three stages of the same process. The left panel shows the process as it looks at the beginning, before any tasks finish. The graphs in the upper panel are colored to show the most likely critical path at the outset, which happens to be along the top branch in this case. The plot below presents the initial distribution of total end-to-end completion times for the entire process, after many repetitions. The jagged blue curve presents the results of the Monte Carlo simulations, and the smooth orange curve presents analytical solutions using the new methods that we will introduce in Chapter 5.

Two factors cause the estimated distribution of completion times to shift during execution. The first is that, after a task is completed, that task no longer has uncertainty in its completion time. It is now history, and all of the uncertainty lies in the unfinished tasks. The second is that, with these changes, the most probable critical path might shift. The example in Figure 3.2 shows both of these effects.

The middle panel shows the process after tasks A and B have been completed. In this instance, task B has been completed earlier than expected. As a result, the overall end-to-end time shrinks (moves to the left), and the critical path shifts to the lower branch. The right panel shows the process after tasks A, B, and C have been completed. In this instance, task C has been completed later than expected. Because this task is now on the critical path, it delays the entire process, as shown in the plot.

This kind of analysis can help project managers consider "what if" scenarios and visualize potential pitfalls as the project progresses. We may also use it to perform sensitivity analyses on individual nodes to understand which tasks have the greatest potential to derail the project. It can also give insight into how the critical path may change.

Because the critical path can shift, we should report not just the most likely critical path but also the probability that each node will fall on the critical path. We define this probability as the *criticality* or *criticality index* of the task or node.[30] In the Monte Carlo method used here, we estimate the criticality of each node by running the graph many times and observing how frequently the node lies on the critical path.

Topological Considerations

Of interest to a project manager are decisions that could help reduce the risk of schedule slippage. In this section, we consider how the structuring of the task dependencies affects the schedule risk. This structuring is called the *topology* of a graph, and, to the degree to which a project manager can control the topology, the manager can mitigate risk.

Idealized Graphs

One option a project manager may have is to remove policy constraints to allow certain work to be done in parallel. To investigate this option, we consider a series of eight graphs constructed by rearranging the same 128 nodes.[31] Initially, all tasks are joined in a single chain. We then split the chain repeatedly, allowing increasing parallelization, until each node lies in its own independent "chain." The result is the set of eight idealized graphs, depicted in part in Figure 3.3, consisting of 1, 2, 4, 8, 16, 32, 64, and 128 separate chains.

Figure 3.3. Splitting a Task Chain into Parallel Development Paths

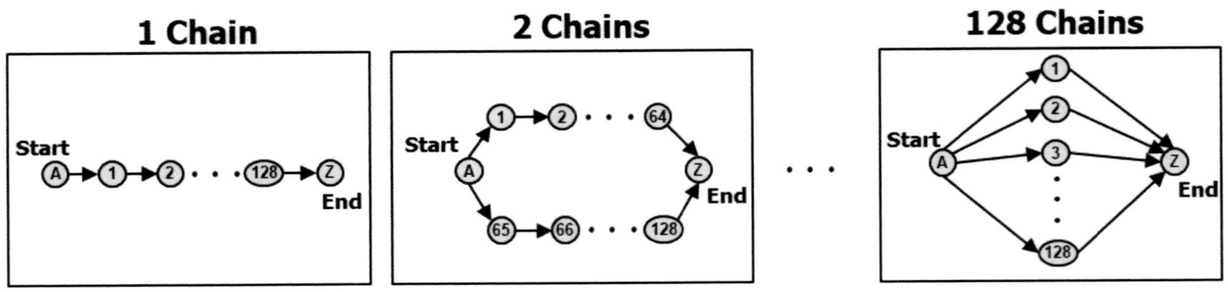

NOTE: All nodes are identical, with most likely completion time of 1, min/max ± 50 percent.

---

[30] Both terms are used in the literature.

[31] Each graph also has a dummy start node and end node with zero completion time, as described in Chapter 1.

Intuitively, of course, we expect that cutting the number of sequential tasks in half repeatedly should reduce the end-to-end time in the same way. And if we plot the raw values of the average end-to-end time for the entire process, we see precisely this in Figure 3.4.

Figure 3.4. End-to-End Completion Times for Different Parallelization Schemes

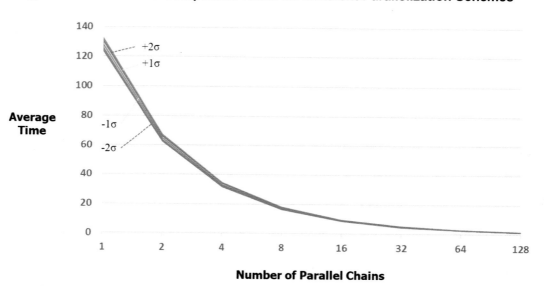

NOTE: The upper green band shows two standard deviations above the mean; the upper yellow band shows one standard deviation above the mean; the lower yellow band shows one standard deviation below the mean; and the lower green band shows two standard deviations below the mean. Each graph was run for 100,000 repetitions.

Closer examination of the result, however, reveals that the reduction in time is not as large as might be expected. In Figure 3.5, we show the same data, but we have now normalized (or normed) the result by dividing by the number of links in the chain.[32] If the end-to-end time simply decreased in proportion to the number of tasks, we would expect a straight line. But this is not what we find.

---

[32] It may be helpful to think of normalized time as requiring a fixed number of worker-hours to complete each task, instead of a fixed number of hours. For a given number of workers, then, dividing the work between serial and parallel topologies has nontrivial effects on the process as a whole.

Figure 3.5. Normed Times for Different Parallelization Schemes

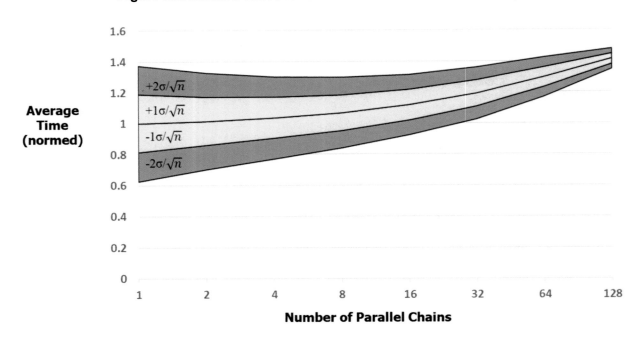

NOTE: The upper green band shows two standard deviations above the mean; the upper yellow band shows one standard deviation above the mean; the lower yellow band shows one standard deviation below the mean; and the lower green band shows two standard deviations below the mean. Each graph was run for 100,000 repetitions.

In Figure 3.5, we see that, as the development path is split into more and more parallel chains, the normed average time grows nonlinearly and the normed variance shrinks.[33] This graph shows the more subtle effect of chaining: As the number of parallel chains grows, the likelihood that one of these paths will take an unusually long time to complete increases. The end-to-end completion time is, in a sense, held hostage by an unfortunate stroke of bad luck in node duration along some chain, making it the worst. Meanwhile, tasks performed mostly in serial have the advantage that a delay early in the process can be remedied by a speedup later down the line, at the cost of the possibility for multiple delays to pile on top of each other.

This analysis shows that while parallelizing development chains can theoretically reduce the overall time, it does not do so as much as one would expect, because the likelihood of having at least one bad chain becomes higher the more chains there are. Thus, there is a trade-off: Parallelizing development reduces the variance more than expected but also increases the end-to-end time more than expected. Thus, although working tasks in parallel remains desirable, the project manager's expectations for time savings should be tempered. This issue is compounded when a project has multiple sets of subprocesses proceeding in parallel.

---

[33] *Normed average* means divided by the expected average; *normed variance* means divided by the expected variance. This normalization is necessary to make an apples-to-apples comparison among chains with different numbers of nodes.

Randomly Constructed Graphs

The previous section used eight idealized graphs, but a similar effect can be seen in less contrived graphs. In this section, we examine an ensemble of randomly constructed graphs that have a topology called an *Erdős-Rényi topology*. Such graphs are a rich field of mathematics, and there are many methods for generating them. Here, to generate a random graph with a certain number of nodes and a certain number of links, we follow a simple procedure: We begin with a maximally connected graph that has the desired number of nodes and remove links randomly until the desired number of links remain.[34] In following this procedure, if any link removal would cause the graph to be disconnected, we instead redo that random draw to ensure that the graph remains connected.[35]

We can then do the same analysis as before: assume all nodes are identical—with the same average, minimum, and maximum completion times—and see how the normed average and normed standard deviation vary. Figure 3.6 shows the results of 1,280 random graphs, each with 128 nodes and between 170 and 560 links. The average and standard deviation of the completion times were computed for 1,000 Monte Carlo iterations of each graph. (As in the previous section, the most likely time is one unit for each node, minimum/maximum ±50 percent.)

---

[34] For a DAG with n nodes with a common start point and a common end point—the type we use throughout this report—the minimal number of links is $n-1$ and the maximal number of links is $\frac{n(n-1)}{2}$. The maximally connected case occurs when every node is linked directly to every other node that follows it in topological order.

[35] In practice, the time required to generate the graphs was negligible compared with the runtime for analyzing them. The bigger concern is that, given the constraints of a common start point and a common end point, it is possible to generate a graph that has more than the minimum number of links but cannot be reduced further, because any removal would disconnect the graph. All of the chained graphs shown in the previous section are irreducible in this manner. To avoid this problem, we look at random graphs that retain significantly more links than the minimum number.

Figure 3.6. Completion Time Trade-Offs for Random Graph

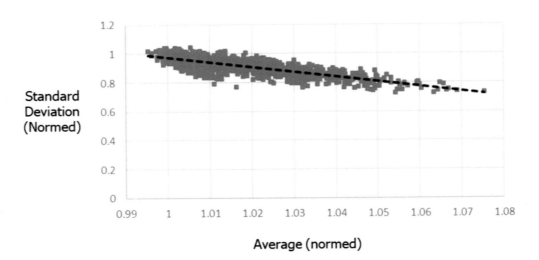

We see in Figure 3.6 the same trade-off between completion time variance and average, though it is more subtle. In this case, the presence of interlinking paths makes it less likely that a single chain will dominate, because the chains are no longer independent.

Scale-Free Graphs

A related aspect of topology concerns the likelihood that a node lies on the critical path—the criticality of that task or node. Nodes with low criticality are unlikely to have a large impact on the schedule. Because we are concerned primarily with the overall project schedule, the primary concern of the project scheduler is to understand the topology of the critical and near-critical paths. We expect a trade-off: Tasks with high criticality that lie in series lead to more uncertainty in the overall completion time (as delays and early finishes add together), whereas parallel chains of tasks lead to less uncertainty but take longer to complete, because it is more likely that at least one chain will be delayed, as described previously.

To investigate this trade-off in more-realistic graphs, we performed another set of Monte Carlo simulations. Many graphs of real-world processes have degree distributions with a fat tail,[36] i.e., most nodes are connected to only a few other nodes, but some are connected to many others. Where the degree distribution is described by a power law,[37] such graphs are commonly referred to as *scale-free*. For these simulations, we generated 500 random scale-free directed

---

[36] The *degree* of a node is the number of links that it is connected to. The *degree distribution* for the whole graph is the probability distribution of these degrees. So, if n_k is the number of nodes with k connections and n is the total number of nodes in the graph, the probability that a node in the graph will have k connections is n_k/n.

[37] The probability of a number of links k decreases by a power function as k increases.

graphs with 500 nodes, each using the Bollobás-Borgs-Chayes-Riordan algorithm.[38] To each graph, we added start and end nodes and used the symmetric beta distribution with most likely time of 1, minimum/maximum ±100 percent. (A larger variance makes it easier to see certain effects.) Each graph was iterated 150 times, and the longest path between start and end nodes was determined. From these calculations, we found approximations for the criticality of each node, and the distribution of the longest paths was fit with a normal distribution to find the mean and variance. The main result is shown in Figure 3.7.

Figure 3.7. Effects of Multiple Near-Critical Paths on Project Completion Time

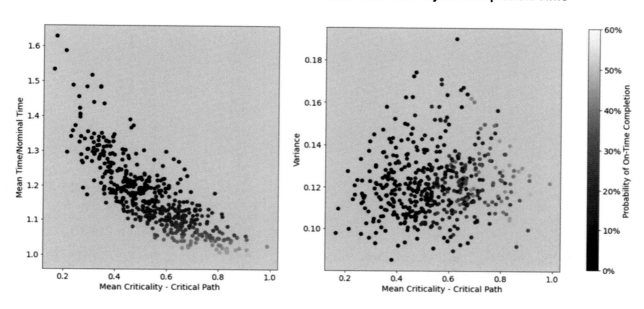

NOTE: The left panel shows the ratio of mean completion time to most likely ("nominal") completion time versus the mean criticality of nodes on the longest nominal path between start and end nodes. Each point corresponds to a different graph. The right panel shows the same data, except here the y-axis represents the variance of the normalized completion time.

For the purpose of visualization, we colored each graph by the probability of completion ahead of schedule: Black points indicate graphs that are almost certain to complete behind schedule, while yellow and red points are more likely to finish on time or early.[39] In Figure 3.7, we see a strong correlation between the mean criticality of the nodes along the longest path and the mean time to complete the traversal of the graph—in other words, graphs with a single,

---

[38] In following this procedure, we drew random values for each of the α, β, and γ parameters that control the graph construction to span a variety of topologies. See Béla Bollobás, Christian Borgs, Jennifer Chayes, and Oliver Riordan, "Directed Scale-Free Graphs," *Proceedings of the Fourteenth Annual ACM-SIAM Symposium on Discrete Algorithms*, 2003; and Béla Bollobás, Christian Borgs, Jennifer Chayes, and Oliver Riordan, "Percolation on Dense Graph Sequences," *Annals of Probability*, Vol. 38, No. 1, January 2010.

[39] This distinction is defined by whether the mean is more or less than 1 standard deviation above unity.

dominant longest path are much more likely to complete on time than graphs with several parallel paths that may each be the critical path.

This result is consistent with the trade-off described in Figure 3.5. However, although the correlation between mean criticality has a correlation coefficient of 0.08, it is much less than the correlation with the mean (coefficient of -0.79).[40] This result means that, for scale-free graphs, the project manager can reduce the overall completion time with only a marginal increase in the corresponding risk by placing important tasks in series with one another. We caution that this result may not be robust to different assumptions: Each of the nodes considered here had identical beta distribution parameters, whereas a real graph is more likely to have a mix of different parameters that may be correlated with degree distribution.

---

[40] We note that the majority of the other metrics related to graph topology we investigated had little impact on the mean or variance. The slope of the power law describing the degree distribution, for instance, had a correlation coefficient of -0.03 when compared with the normalized completion mean, and 0.12 with the normalized variance. Nodes with high in-degree, although likely to start behind schedule, are relatively rare compared with nodes with smaller degree and therefore do not contribute strongly to schedule risk.

4. Resource Constraints

In Chapter 2, we showed how the principle of reducing the risk of rework could be used to generate task dependency graphs. In Chapter 3, we applied probabilistic completion time information to each task node and showed how the critical path and graph topology affect the end-to-end completion times for the entire process. In doing so, however, we implicitly assumed that work begins on each task as soon as possible—i.e., as soon as all predecessor tasks are completed. This assumption ignores resource constraints. In this chapter, we add resource requirements to each node and show how resource constraints can affect scheduling options.

Fixed Resource Requirements

The simplest kind of resource requirements are fixed requirements, in which it is known in advance how many resources are needed to complete each task. Resources could be expressed in various ways, including person-hours or dollars. As an example, consider the graph depicted in Figure 4.1. This graph consists of five nodes or tasks, labeled A through E. The most likely completion time for each task is displayed above the node in green. The resource requirement for each task is displayed below the node in red. In this simple example, the time and resources do not vary. Each task takes precisely as long as indicated—no more, no less—and, for simplicity, each task requires exactly one unit of labor, e.g., one worker. (As usual, we also include dummy start and end nodes, which require neither time nor resources.)

Figure 4.1. Simple Example of Fixed Time and Resource Constraints

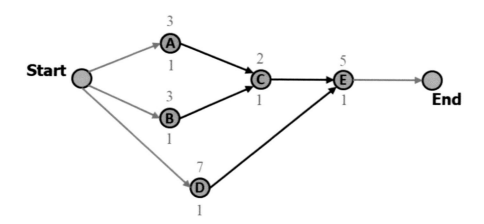

NOTE: Time requirements are displayed above each node in green; resource requirements are displayed below each node in red.

25

It is evident that the lower branch of the graph (D, E) corresponds to the most likely critical path, as it takes 12 time units to complete, whereas either of the upper paths (starting with A or B) takes only 10 time units. However, to complete the entire project within the critical time requires tasks A, B, and D to begin simultaneously, which requires 3 labor units to be spent at the same time. If only two workers are available, this cannot be done, and the critical path cannot be achieved.[41]

Figure 4.2 compares the constrained and unconstrained cases. The black bars show the labor use over time when there are no resource constraints. The total end-to-end completion time is 12 time units, as expected. The gray bars show what happens if there is a resource cap of two workers. Starting along the critical path, we may assign one worker to task D while the other completes tasks A, B, and C. The first worker finishes task D at time 7 but must wait to begin work on the final task until time 8, when task C is complete. The total end-to-end completion time becomes 13 time units.

Figure 4.2. Effects of a Resource Constraint

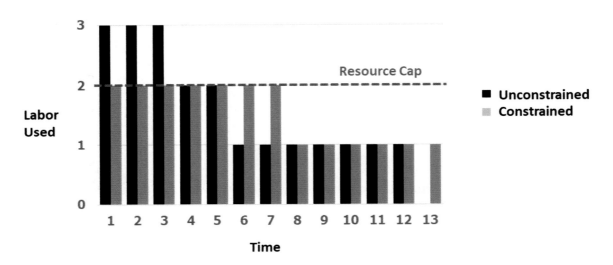

In other words, if there are no constraints, or if additional labor can be obtained early in the project, the entire project will remain on schedule—but if this is not the case, the schedule must slip. This example is a quantitative illustration of the qualitatively obvious point that the addition of a resource cap can delay the overall project beyond the timeline indicated by the critical path,

---

[41] The resource-constrained project scheduling problem (RCPSP) is a known, difficult (called *non-polynomial-hard*) problem. Several authors have studied it and used heuristic methods or mixed-integer linear programming (MILP) to solve it and variants of the problem. See, for example, Juan D. Palacio and Olga L. Larrea, "A Lexicographic Approach to the Robust Resource-Constrained Project Scheduling Problem," *International Transactions in Operational Research*, Vol. 24, No. 1/2, January–March 2017.

even though the total labor consumption does not change. It also leads us to consider the concept of *slack* in the system.

Slack

In Chapter 3, we implicitly assumed that each task started as early as possible. However, while the tasks on the critical path must begin as soon as possible—if there is any delay along the critical path, the overall schedule will be negatively affected—tasks that are off the critical path may sometimes be delayed without affecting the overall schedule. Unlike the taut critical path, off-critical paths have *slack*, meaning that there is some freedom to slide those tasks earlier or later in the schedule without affecting the end-to-end completion time of the entire project. The total amount of slack in the system is an important way of understanding how robust the critical path is to resource constraints.

To characterize the amount of slack in the project, we compare two edge cases. In the first, we follow the standard approach in which all tasks are scheduled to begin as early as possible. In the second, all tasks are scheduled to begin as late as possible *while still adhering to the overall end-to-end time dictated by the critical path*. In general, the first approach is preferred when the deadline is of paramount importance, while the second may be selected to meet resource constraints.[42]

For an individual task, the *slack* is generally defined as the difference between the latest and earliest possible start times for that task, assuming no other tasks are delayed.[43] However, the concept of slack in the entire system is more complicated, as tasks along the same development path can "share" the same slack. Figure 4.3 shows the overall schedule, comparing scheduling tasks as early as possible (the same black bars as in Figure 4.2) with scheduling them as late as possible.

---

[42] Krzysztof S. Targiel, Maciej Nowak, and Tadeusz Trzaskalik, "Scheduling Non-Critical Activities Using Multicriteria Approach," *Central European Journal of Operations Research*, Vol. 26, No. 3, September 2018.

[43] Frederick S. Hillier and Gerald J. Lieberman, *Introduction to Operations Research*, 10th ed., New York: McGraw-Hill Education, 2014.

Figure 4.3. Slack in the System

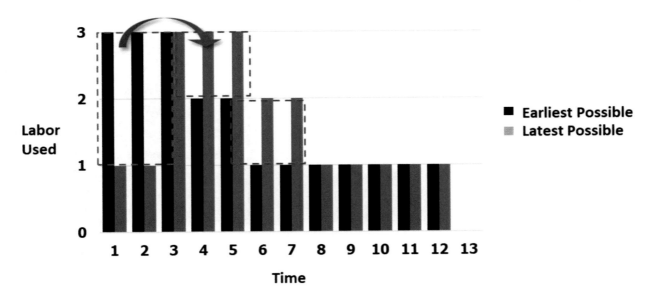

Moving from the earliest case to the latest case, much of the labor allocation is the same, but the labor in the red box shifts to the purple boxes in Figure 4.3. The total amount that can be shifted consists of 2 labor units over 3 time periods. Intuitively, this means that there are 6 total units of slack in the system. With this graph, there is flexibility to shift the peak labor demand from the beginning toward the middle of the project, but the total amount of labor does not change, and the peak remains at the same height. A more interesting graph, however, offers more flexibility.

Variable Resource Requirements

Now we consider a more complex example in which both time and resource requirements vary stochastically. Figure 4.4 shows a slightly more complicated graph, showing the minimum and maximum completion times for each task above the corresponding node in green and the minimum and maximum labor requirements for each task below the corresponding node in red. There is no standard method for extending PERT in this way. For consistency, we use the same beta distribution for the variability in labor as we do for the variability in time.

Figure 4.4. More Complex Example of Time and Resource Constraints

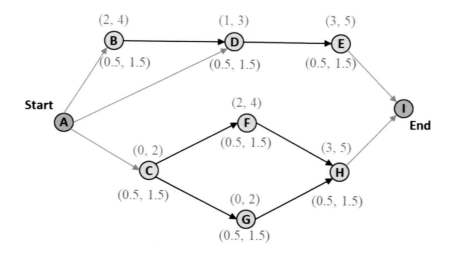

This graph has been constructed so that the effective critical path may shift from the top to the bottom, depending on the actual completion times of each task. The criticality of each node is shown in Figure 4.5. The criticality map is simple: The percentage is 0, 14, or 86 percent for any given node. As expected, the most-likely critical path is the top branch, with the second-most-likely being the middle branch. The bottom branch (i.e., the path that includes task G) is never critical.

Figure 4.5. Criticality of Nodes in the More Complex Example

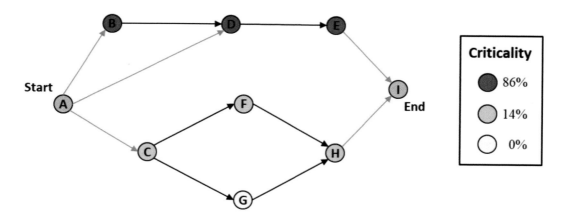

We now look at the same two edge cases. Figure 4.6 shows the labor allocation graph. In black, we see the standard case, in which all tasks are scheduled to begin as early as possible. In red, we see the other edge case, in which all tasks are scheduled to end as late as the critical path will allow.[44]

Figure 4.6. Edge Cases in the Scheduling Problem

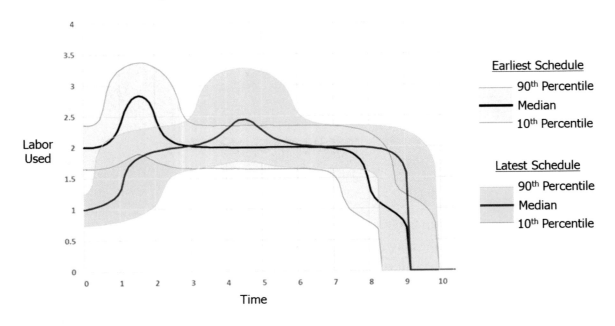

By shifting from earliest-start to latest-end scheduling, the overall schedule is unchanged, but the labor allocation is somewhat different. The peak allocation is somewhat diminished, and the labor is more evenly spread over time. Of course, the nature of the shift depends entirely on the nature of the underlying graph. It also raises the following questions: What about an intermediate case? What if the tasks are scheduled somewhere in the middle of the slack intervals rather than at the beginning or end of the allowed intervals? How can tasks be arranged so as to minimize surges in resource demand?

We are not aware of any algorithm that can solve this general scheduling problem without a brute-force approach, and we do not attempt to provide one here. There are various existing heuristic approaches, but none incorporate the kinds of uncertainties described here. Although a heuristic solution is not guaranteed to be optimal, we note that it might provide a good result that is far more computationally convenient to calculate.

We suggest a heuristic that prioritizes the nodes that are likely to lie along the critical path in the graph. It is clear how this type of prioritization can favorably break the degeneracy between different topological orders in the simple example introduced at the beginning of this chapter: The longest weighted path through the graph is (Start, D, E). In a stochastic scenario, node D is

---

[44] Here, we use the symmetric beta distribution for time, using the optimistic and pessimistic times (the pair shown in green above each task) from Figure 4.4. Likewise, we use a symmetric beta distribution for resource consumption, using the optimistic and pessimistic resource consumptions (the pair shown in red below each task) from Figure 4.4.

likely to lie on the critical path, so we should prioritize topological orders that place D earlier in the process—in other words, the criticality heuristic suggests that the topological order (Start, D, A, B, C, E) is superior to (Start, A, B, C, D, E). If the scheduling prioritization is updated regularly as tasks are completed, we expect the results for the entire process to be close to (though not guaranteed to be) optimal.

Given the complexity of real-world projects and their resulting graph representations, the analytic approaches described in the next chapter are likely necessary to compute the criticality in a reasonable amount of computation time, which will then inform the prioritized schedule under resource constraints in the manner described here.

5. Analytic Approach for Calculating the Criticality Index

When we leave the world of small graphs, the runtime for the Monte Carlo method can become prohibitive because the number of possible paths grows exponentially with size.[45] This chapter formally derives the mathematics for computing the criticality index on large graphs. The method presented is novel and decreases computational times by orders of magnitude.

Figure 5.1 shows an example of how runtime grows for graphs with an increasingly large number of links (edges) on a typical personal computer.[46] Unfortunately, many important topological features emerge only in relatively large graphs, and many real graphs are quite large. And, as noted previously, the task dependency graph for a large program, such as GBSD, is likely to include tens of thousands of nodes, if not more, with perhaps millions of links. More-advanced mathematical techniques allow us to design a faster numerical method for computing the criticality index for large graphs. These techniques also allow us to prove convergence and quantify convergence in terms of the regularity of PDFs—meaning that we can be confident that the method will always succeed, and we are not limited by the number of Monte Carlo runs we can make.

Figure 5.1. Runtime Problems for the Monte Carlo Method

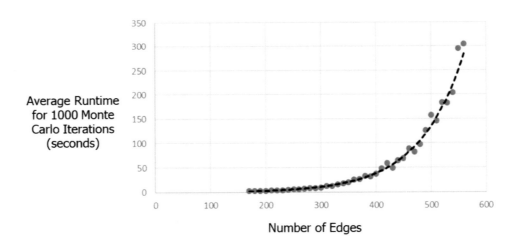

NOTE: Each Erdős-Rényi graph contained 128 nodes and had a varying number of links. Results are shown for a Dell Latitude E6440, Intel i7-4610M, 3 GHz; 8-GB RAM.

---

[45] The number of possible paths to traverse a fully connected graph of n nodes goes as 2^n.

[46] This figure shows the runtime to analyze random graphs.

Approach

In this chapter, we present an analytic approach for assessing the criticality of activities in a PERT graph using spectral methods. This application of spectral methods permits the analysis of very large graphs. The basic approach is to transform the problem into a different space, perform the calculations efficiently in the transformed space, and then invert the transformation back to the original space. So long as the effort to do the transformations is small compared with the savings of computing in the transformed space, the method is efficient. The concept is shown schematically in Figure 5.2.

Figure 5.2. Calculations Using Transformations

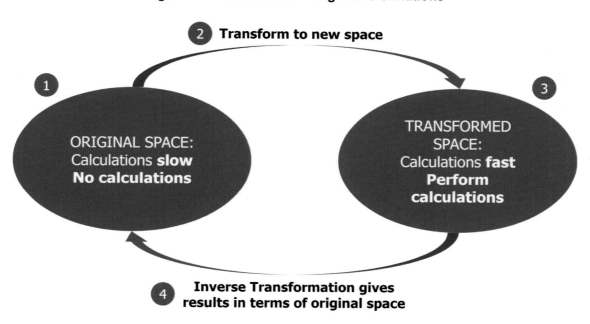

A simple analogy of this approach is how multiplication was often done prior to modern computers. To multiply a sequence of numbers to high precision, the steps were to calculate the logarithm of each, add them in the transformed space of the logarithm, and, finally, invert the transformation with an exponential to get the correct answer. The operation of multiplication is reduced to a summation. As long as the transformation is not burdensome, the calculation is simpler. In the days before digital calculators and computers, slide rules performed the logarithm graphically, making multiplication a simple problem of addition.

In a similar manner, the Fourier transform (and, by extension, a Fourier series representation) allows for the simplification of several mathematical operations, including differentiation, integration, and convolution. In particular, once a function is transformed via the Fourier transform, differentiation and integration are performed by a simple multiplication of the frequency parameter while convolution corresponds to multiplication of functions in the transformed space. Additionally, Fourier series representations exhibit rapid convergence, thus allowing for a reduction in the information needed for calculations.[47] These properties have been exploited over the past several decades to develop efficient methods for solving differential equations that arise in a range of modeling contexts.[48] In this chapter, we use similar techniques to develop an efficient numerical method for performing criticality index calculations.

To use these transform methods, we develop a semi-analytical approach that allows the criticality index to be expressed in terms of a series of differentiations, integrations, and convolutions of the underlying PDFs for each node of a PERT network. Following this, we show how the semi-analytical construction can be used to develop two numerical methods for calculating the criticality index. The first method uses a standard discretization of the underlying PDFs, while the second method uses a discretization in a transformed space. By comparing these two approaches, we show the benefit of the latter approach for calculating the criticality index.

The remaining parts of this chapter formally derive these results.

Formalism and Notation

We will start with a few formal definitions to fix terminology and notation:

- Let $V := \{v_1, \dots, v_n\}$ be a finite set of elements, which will be referred to as *nodes* (or *vertices*), and let $V \otimes V$ denote the set of ordered pairs (v_i, v_j) of elements in V.
- A relation on the set V is any subset $E \subset V \otimes V$. The relation E is symmetric if $(v_i, v_j) \in E$ implies $(v_j, v_i) \in E$ for all $i, j \in \{1, \dots n\}$. The relation E is said to be antireflexive if $(v_i, v_j) \in E$ implies $v_i \neq v_j$ for all $i, j \in \{1, \dots, n\}$.
- A graph $G = (V, E)$ is an ordered pair where V is a set of nodes and E is a set of symmetric antireflexive relations on V (in particular, $E \subset \{\{v_1, v_2\} \mid v_1, v_2 \in V, v_1 \neq v_2\}$). An element of the set E is referred to as a *link* (or an *edge*).
- A link $e = \{e_1, e_2\} \in E$ is said to be adjacent to the node $v \in V$ if $e_1 = v$ or $e_2 = v$.
- Two links $e = \{e_1, e_2\}, f = \{f_1, f_2\} \in E$ are said to be adjacent if $e_i = f_j$ for any $i, j \in \{1, 2\}$.
- We use $v_j \in \mathrm{Pd}(v_i)$ to denote that v_j is a predecessor of v_i. Similarly, we use $v_j \in \mathrm{Sc}(v_i)$ to denote that v_j is a successor of v_i.

The set of nodes and links in a graph G may be denoted by V_G and E_G, respectively, to emphasize the association with G.

---

[47] Elias M. Stein, *Harmonic Analysis: Real-Variable Methods, Orthogonality, and Oscillatory Integrals*, Princeton, N.J.: Princeton University Press, 1993.

[48] John P. Boyd, *Chebyshev and Fourier Spectral Methods*, 2nd ed., Mineola, New York: Dover Publications, Inc., 2000.

Criticality Index

If each node in a DAG has a specified weight (such as completion time), then the longest path can be computed deterministically using a depth-first search algorithm or the algorithm described in Appendix B. Finding node criticality from the longest path is then trivial. Nodes that are on the longest path have a criticality of 1, and other nodes have a criticality of 0. However, in the case in which the task duration is drawn from a probability distribution, the critical path may change because of the variation in the time for each draw. Therefore, we need a method for finding the probability of each path being critical.

The idea of studying which paths are likely to be critical was introduced by Richard Van Slyke.[49] J. J. Martin distinguished the concept of the criticality of a path from the concept of the criticality of a node.[50] Subsequently, there have been several developments that improve the efficiency of computing the criticality index of each node.[51] In this section, we introduce a new approach to calculate the criticality index. In this approach, the calculation proceeds in two stages—first, computing the PDFs of each node's completion time by propagating uncertainty forward through the graph and, second, calculating the probability of each node lying on the critical path by moving backward through the graph. Additionally, we present two methods for performing the calculations—one by discretizing the relevant PDFs and using a standard trapezoidal rule for integration, and the other by using a Fourier spectral approximation with a Chebyshev basis.

Calculation of the Criticality Index

Let $\mathcal{P}$ denote the set of all paths through a given PERT graph, and let $D(p_i)$ denote the duration of path $p_i \in \mathcal{P}$. The duration of a path can be written in terms of the duration of the individual nodes that comprise the path as

$$D(p_i) = \sum_{v_j \in p_i} d(v_j),\tag{5.1}$$

where $d(v_j)$ denotes the duration for the node v_j, which is on the path p_i. If each of the $d(v_j)$ is stochastic and follows the distribution of a random variable (say X_j) with a well-defined PDF, then the possible values of $D(p_i)$ will also follow the distribution of a random variable. This

[49] Van Slyke, 1963.

[50] J. J. Martin, "Distribution of the Time Through a Directed, Acyclic Network," *Operations Research*, Vol. 13, No. 1, 1965.

[51] Bajis M. Dodin, "Approximating the Distribution Functions in Stochastic Networks," *Computers & Operations Research*, Vol. 12, No. 3, 1985; Bajis M. Dodin and Salah E. Elmaghraby, "Approximating the Criticality Indices of the Activities in PERT Networks," *Management Science*, Vol. 31, No. 2, February 1985; and S. M. T. Fatemi Ghomi and E. Teimouri, "Path Critical Index and Activity Critical Index in PERT Networks," *European Journal of Operational Research*, Vol. 141, No. 1, August 2002.

follows from the fact that the sum of two random variables is again a random variable.[52] Furthermore, the density function for the random variable corresponding to $D(p_i)$ will be the convolution of the PDFs of the respective X_j for each node that lies on the path $p_i \in \mathcal{P}$.

The criticality of a path is the probability that its duration is as long as or longer than every other path in the PERT graph. That is, the criticality of a path $p_i \in \mathcal{P}$ is given by

$$C_{\text{Path}}(p_i) = P[D(p_i) \geq D(p_j), \forall\, p_j \in \mathcal{P}, i \neq j]. \tag{5.2}$$

The criticality index of a node (an activity or task) is often a more useful measure for determining which nodes are likely to lie on the critical path because nodes may lie on more than one path. Node criticality is defined by summing the criticalities of all of the paths on which the node lies. That is,

$$C_{\text{Node}}(v_j) = \sum_{\{p_i \in \mathcal{P}\,:\, v_j \in p_i\}} C_{\text{Path}}(p_i). \tag{5.3}$$

Both $C_{\text{Path}}: \mathcal{P} \to [0,1]$ and $C_{\text{Node}}: V_G \to [0,1]$ are well-defined functions that assign probabilities on the interval $[0,1]$ for each path in $\mathcal{P}$ and node in V_G, respectively.

In the discussion that follows, we present a semi-analytical method for deriving the criticality index of a node. This method proceeds in two stages. First, we calculate the cumulative time the process will take to complete the activity at each node v_i along the "longest" path that leads to v_i. This will be referred to as the *latest cumulative time* (LCT) of the process up to and including node v_i. It is important to distinguish the LCT from the duration associated with the node itself. The latter is the PDF given by the 4-parameter beta distribution and only depends on the nature of the activity that takes place at that node. The calculation of the LCT for each node in the graph proceeds iteratively along any given topological ordering and begins with the start node and terminates at the finish node. Second, we calculate the probability that each node v_i is critical to its successors and, hence, the criticality of the node. By *critical to its successors*, we mean the probability that the node v_i completes at the latest time among all of the predecessors of v_j. This set of calculations depends on the LCT of each node calculated in the first step. It begins at the end node and proceeds in reverse topological order to the start node.

In general, the duration of each node in the PERT graph is stochastic but may be drawn from a known distribution. In this section, we assume that the duration of each node v_j follows a distribution with known cumulative distribution function (CDF), denoted by $F_{X_i}: \Omega_i \to [0,1]$, where $\Omega_i \subset \mathbb{R}$. We assume that F_{X_i} is continuously differentiable (i.e., $F_{X_i} \in C^1(\Omega_i)$) and that a

---

[52] Walter Rudin, *Real and Complex Analysis*, 3rd ed., New York: McGraw-Hill, 1987.

corresponding PDF exists, denoted by $f_{X_i}: \Omega_i \to \mathbb{R}^+$.[53] These conditions hold for the 4-parameter beta distribution, but the method can be extended to other distributions.

We now begin with the first of the two stages for calculating the criticality index—finding the LCT for each node, beginning at the start node and proceeding to the finish node along a topological ordering. Because the duration of each node is stochastically determined, it is not immediately apparent which path will be the critical path. However, by computing the *maximum order statistic* (MOS) over all predecessors leading to node v_i,[54] we can determine the probability that a given predecessor lies on the critical path and hence the probability that v_i lies on the critical path.

Let $f_{v_i}^{MOS}$ denote the MOS for the completion time of the predecessors of node v_i. The PDF for the MOS is written in terms of the LCT of its predecessors as follows (see Appendix D for a derivation of this equation):

$$f_{v_i}^{MOS}(t) = \frac{\partial}{\partial t}\left[\prod_{v_j \in \mathrm{Pd}(v_i)}\left(\int_0^t f_{v_j}^{LCT}(t')\, \mathrm{d}t'\right)\right]. \tag{5.4}$$

Note that the LCTs $\left\{f_{v_j}^{LCT}\right\}_{v_j \in V_G}$ in general are assumed to be independent but not identically distributed, since each depends on the paths that lead to the node $v_j \in V_G$ for each $j \in \{1, \ldots, N\}$.

The time that a node is completed is the sum of the node's intrinsic duration time and the MOS of its predecessors. Because both are probability distributions, we convolve the two distributions to find the probability distribution of their sum. The convolution is

$$f_{v_i}^{LCT}(t) = \left(f_{v_i} * f_{v_i}^{MOS}\right)(t) = \int_{-\infty}^{\infty} f_{v_i}(t')\, f_{v_i}^{MOS}(t' - t)\, 1_{\{t \in \Omega_i\}}\, \mathrm{d}t'. \tag{5.5}$$

Once the cumulative time is calculated for node v_j, it is used to compute the cumulative duration for each of its successors using the same process outlined above. Because it is a dummy

---

[53] Note that this regularity condition can be relaxed. In particular, if F_{X_i} is absolutely continuous (see Appendix D for a definition of *absolute continuity*) on Ω, then $f_{X_i}(x)$ exists almost everywhere and is the Radon-Nikodym derivative of $F_{X_i}(x)$ with the Lebesgue measure.

[54] Let $\{X_i\}_{i=1}^n$ be a set of random variables. The order statistics, denoted by $X_{(1)}, \ldots, X_{(n)}$, are defined by sorting the values of X_i in increasing order. The MOS is given by $X_{(n)} = \max\{X_1, \ldots, X_n\}$. The MOS is itself a random variable with an associated CDF and PDF.

node, the start node is assumed to occur at a fixed (deterministic) period of time; hence, its probability distribution can be represented by the Dirac delta generalized function $\delta_{t_0}(t)$.[55]

The previous calculations give the LCT of each of the nodes. Next, we turn our attention to the calculation of the criticality index. For this calculation, we work backward from the finish node, which has a criticality of 1. (Since all paths from start to finish pass through it, it must be on the critical path.) Let $P_C(v_i, v_j)$ for $v_j \in Sc(v_i)$ denote the probability that node v_i is critical to each of its successors—that is, the probability that the node v_i completes at the latest time among all of the predecessors of v_j. The criticality of node v_i can be written as

$$C_{\text{Node}}(v_i) = \sum_{v_j \in Sc(v_i)} C_{\text{Node}}(v_j)\, P_C(v_i, v_j). \tag{5.6}$$

The probability $P_C(v_i, v_j)$ that a given node v_i will be the critical predecessor for any node v_j can be found directly from the completion time PDFs. This is done by considering each predecessor node in turn in comparison with the MOS for all of the other predecessors. With this "one versus rest" strategy, we need to perform a two-dimensional integral for each node at the cost of recomputing the MOS for all of the predecessors of v_j:

$$P_C(v_i, v_j) = \int_0^\infty \int_{t_j}^\infty f_{v_i}^{LCT}(t_i)\, f_{v_{j \neq i}}^{MOS}(t_j)\, dt_i\, dt_j. \tag{5.7}$$

Once these calculations are finished, the node's completion time PDF is no longer uncertain and therefore collapses into a Dirac delta distribution. This reduces uncertainty in the successor nodes, and the new uncertainties propagate forward along the paths in the graph. Each *time* a node is completed, we recompute the PDFs and the criticality indices to account for the new information.

This analytic computation of the critical path gives the necessary mathematical description to enable the transformation and integration.

Numerical Methods for Calculating the Critical Path

We now present two numerical approaches for implementing the semi-analytical method presented above, one of which we call *trapezoidal integration* and the other the *spectral method*. Both approaches are performed by deriving approximations of the functions in the semi-analytical method. However, the method of approximation differs between the two techniques.

---

[55] For a formal definition of a generalized function, see Lars Hörmander, *The Analysis of Linear Partial Differential Operators I: Distribution Theory and Fourier Analysis*, Berlin: Springer, 2015.

The first approach uses a standard trapezoidal approximation for integration and a second-order finite difference scheme for differentiation. The second method, referred to as the *Chebyshev spectral method*, uses a Fourier series representation with a Chebyshev basis to approximate the underlying PDFs used in the semi-analytical construction. The criticality calculations are then performed in terms of integrals, derivatives, and convolutions of the Chebyshev series representations.

As mentioned at the beginning of this chapter, one motivation for deriving these methods is to improve the efficiency of the criticality computation. One way of measuring the efficiency of an algorithm is to study its rate of convergence, i.e., the rate at which the error for an approximation approaches zero. The rate of convergence for the Monte Carlo method is $O(1/\sqrt{M})$, where M is the number of runs used in the simulation.[56] We will show that the rate of convergence of the method with trapezoidal integration is $O(M^{-2})$, where M is the number of partitions used in the approximation. The rate of convergence of the Chebyshev spectral method is $O(e^{-M})$, where M is the number of terms in the Chebyshev series representation of the PDFs.

In general, the spectral method is faster, but some software packages do not provide strong support to manipulate Chebyshev series, so we present both methods. Figure 5.3 shows the relative rates of convergence of these two methods in comparison with the Monte Carlo simulations. The shallow slope of the Monte Carlo simulations reflects the need for extended computational time, whereas the method using the Chebyshev series converges very rapidly and requires little computational time.

---

[56] Van Slyke, 1963.

Figure 5.3. Relative Rates of Convergence for the Three Methods

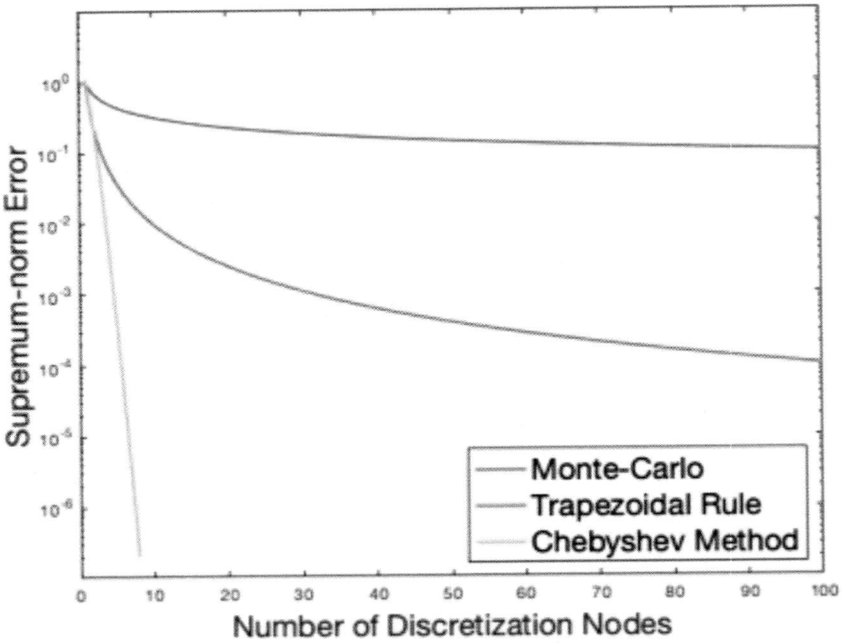

Numerical Approximation with Trapezoidal Integration

We will now show that the calculations for the semi-analytical method can be performed using the trapezoidal rule for integration and a second-order method of differentiation. There are three important distributions that need to be approximated. These are $f_{X_j}^{MOS}$, $f_{X_j}^{LCT}$, and $P_C\left(v_i, v_j\right)$. By approximating each of these with the trapezoidal rule and a second-order differentiation method, we will develop a numerical implementation of the semi-analytical method. Given a specified level of accuracy, this approach is a more efficient method for approximating criticality than the Monte Carlo method.

Approximation of the Density Function for the Maximum Order Statistic

Now we comment on the approximation of f_i^{MOS}. Fix $i \in \{1, \dots, N\}$ and note that the inner integral for each of the $\left\{f_{v_j}^{LCT}\right\}_{v_j \in Pd(v_i)}$ (see Equation 5.4) can be approximated using the cumulative trapezoidal rule. Let $F_{v_j}^M$ denote the cumulative trapezoidal approximation of the CDF of $f_{v_j}^{LCT}$ with M uniformly spaced discretization intervals. The function $F_{v_j}^M$ is defined on the interval $[t_0, t_M]$ pointwise at $\{t_\alpha\}_{1 \leq \alpha \leq M}$ by

$$F_{v_j}^M(t_\alpha) = \frac{1}{2} f_{v_j}^{LCT}(t_0)\, \Delta t \; + \; \sum_{\beta=1}^{\alpha} f_{v_j}^{LCT}\left(t_\beta\right) \Delta t, \tag{5.8}$$

where $\Delta t = t_\alpha - t_{\alpha-1}$, and the Mth term is multiplied by $\frac{1}{2}$. As $M \to \infty$, the error for the trapezoidal approximation decreases as $O(M^{-2})$.[57]

Let $F_{\Pi_{v_i}}^M$ denote the product of the $F_{v_j}^M$ over all $v_j \in Pd(v_i)$. The PDF for the MOS is approximated using a second-order central finite difference scheme as follows:

$$f_{v_i}^{MOS}(t_\alpha) \approx \frac{F_{\Pi_{v_i}}^M(t_\alpha + \Delta t) - 2\,F_{\Pi_{v_i}}^M(t_\alpha) + F_{\Pi_{v_i}}^M(t_\alpha - \Delta t)}{(\Delta t)^2}, \tag{5.9}$$

for $\alpha \in \{1, \dots, M-1\}$, and, again, $\Delta t = t_\alpha - t_{\alpha-1}$. As $M \to \infty$, the error for this approximation decreases as $O(M^{-2})$. Hence, this is a second-order approximation of $f_{v_i}^{MOS}$.

Approximation of the Density Function for the Latest Critical Time

Now we consider $f_{X_i}^{LCT}$. Within each iteration of the latest critical time calculation, $f_{X_i}^{LCT}$ is computed by performing the convolution of f_{X_i} with $f_{X_i}^{MOS}$. We perform this convolution by first transforming f_{X_i} and $f_{X_i}^{MOS}$ using the Fast Fourier Transform. Once transformed, the convolution can be computed efficiently using the Fourier Convolution Theorem.[58] The resulting series is then transformed back using the inverse Fast Fourier Transform.

Approximation of the Density Function for the Critical Predecessor

We will use a midpoint approximation to perform the double integral for $P_C(v_i, v_j)$. Let $\bar{t}$ denote the midpoint of the interval $(t, t + \Delta t)$:

$$\bar{t} = \frac{t + (t + \Delta t)}{2}. \tag{5.10}$$

The double integral for $P_C(v_i, v_j)$ can be approximated as follows:

$$P_C(v_i, v_j) \approx \frac{1}{M^2} \sum_{\alpha=1}^{M} \sum_{\beta=1}^{M} f_{v_i}^{LCT}(\bar{t}_\alpha)\, f_{v_{j \neq i}}^{MOS}(\bar{t}_\beta)\, (\Delta t)^2. \tag{5.11}$$

Again, for large M, this approximation is second-order accurate.

---

[57] Note that we index the Chebyshev basis and numerical approximations with Greek letters to distinguish them from indices used for graph components.

[58] Rudin, 1987.

Numerical Approximation of the Critical Path with the Spectral Method

We will now show that the calculations for the semi-analytical method can be performed with a spectral method. The spectral method proceeds by representing each of the PDFs as a Fourier series. The calculations are then performed directly on the Fourier series representations. There are several benefits to this approach, including two that we mention here. First, the Fourier series of a well-behaved function converges rapidly, which considerably reduces the amount of information that needs to be stored to accurately represent functions and perform calculations. Second, several standard operations can be performed with greater accuracy and lower complexity with a Fourier series representation. In particular, differentiation is reduced to multiplication by a constant, and convolution is performed by summing Fourier coefficients. Taken together, these benefits result in an efficient algorithm for performing the calculations needed to compute the criticality index.

To implement the semi-analytical criticality index method, there are three important distributions that need to be approximated—these are $f_{X_i}^{MOS}$, $f_{X_i}^{LCT}$, and $P_C\left(v_i, v_j\right)$. We will show how each of these can be approximated with a Chebyshev series representation, thus giving a numerical implementation of the method.

The Chebyshev polynomials, denoted by $\{T_\alpha(x)\}_{\alpha=0}^\infty$, are defined by the following two-term recurrence relation:

$$T_0(t) = 1, T_1(t) = t$$
$$T_\alpha(t) = 2t\, T_{\alpha-1}(t) - T_{\alpha-2}(t). \tag{5.12}$$

Figure 5.4 shows the first five Chebyshev polynomials.

Figure 5.4. The First Five Chebyshev Polynomials

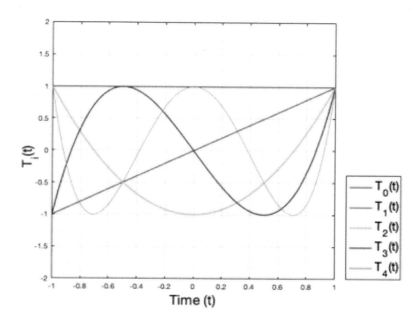

The Chebyshev polynomials form an orthogonal L^2-basis on $[-1,1]$. The Chebyshev polynomials are an ideal basis for constructing a spectral method for two reasons. First, they quasi-minimize the supremum-norm error;[59] therefore, they form a more accurate approximation than the typical cosine and sine bases used for a standard Fourier series. Second, unlike the cosine and sine functions, the Chebyshev polynomials are not periodic, which means that nonperiodic functions can readily be approximated with a Chebyshev series without artificially enforcing periodic boundary conditions.

With the inclusion of each successive basis function in the Chebyshev approximation, the error for the approximation decreases as $O(e^{-M})$, where M is the number of terms in the Chebyshev series.

Initial Projection into Chebyshev Basis

We begin by approximating each density function for the duration of each node, f_{X_i}, with a Chebyshev series. Note that this needs to be done only once for each unique PDF. (If multiple nodes have the same PDF, then the same approximation can be used.) The Mth-order Chebyshev series approximation is given below:

---

[59] For a continuous function, the supremum-norm is the maximum value of the pointwise distance between two functions on an interval. In particular, let $f(t)$ and $g(t)$ be continuous functions defined on a compact interval Ω. The supremum-norm is the maximum value of $|f(t) - g(t)|$ over all $t \in \Omega$. We use the supremum-norm because it provides a stricter bound than all other L^p-norms, given that Ω is compact.

$$f_{X_i}(t) \sim \frac{a_0}{2} + \sum_{\alpha=1}^{M} a_\alpha \, T_\alpha(t), \tag{5.13}$$

where $\{T_\alpha\}_{\alpha=1}^{\infty}$ is the basis consisting of Chebyshev polynomials. The coefficients $\{a_\alpha\}_{\alpha=0}^{M}$ are calculated using the following inner product on $[-1,1]$:

$$a_\alpha = \int_{-1}^{1} w(t) \, f_{X_i}(t) \, T_\alpha(t) \, dt, \tag{5.14}$$

where $w(t)$ is the weight given by

$$w(t) = \frac{1}{\sqrt{1-t^2}}. \tag{5.15}$$

In practice, the coefficients can be efficiently approximated as follows using the trapezoidal rule (with m subdivisions) and the fact that $T_\alpha(\cos\theta) = \cos(\alpha\,\theta)$ for $\theta \in [0,\pi]$:

$$a_\alpha = \frac{2}{m} \sum_{\beta=1}^{m} f_{X_i}(t_\beta) \, T_\alpha(t_\beta) = \frac{2}{m} \sum_{\beta=1}^{m} f_{X_i}\left[\cos\left\{\frac{\pi\left(\beta-\frac{1}{2}\right)}{m}\right\}\right] \cos\left\{\frac{\pi\,\alpha\left(\beta-\frac{1}{2}\right)}{m}\right\}. \tag{5.16}$$

Approximation of the Density Function for the Maximum Order Statistic

Now we comment on the approximation of $f_{X_i}^{MOS}$. Fix $i \in \{1,\ldots,N\}$ and note that each of the $\{f_{X_j}^{LCT}\}_j$, where v_j is a predecessor of v_i, will have a Chebyshev representation of the form

$$f_{X_j}^{LCT}(t) \sim \frac{b_0}{2} + \sum_{\alpha=1}^{M} b_\alpha \, T_\alpha(t). \tag{5.17}$$

The calculation of f_i^{MOS} is expressed in terms of an integral, product, and derivative of the $\{f_{X_j}^{LCT}\}_j$ for $v_j \in \mathrm{Pd}(v_i)$. These operations are performed using standard techniques that can be found in the literature on Chebyshev approximations.[60]

---

[60] Lloyd N. Trefethen, "Approximation Theory and Approximation Practice," *SIAM*, Vol. 164, 2019, chapters 3 and 8; and Theodore J. Rivlin, *The Chebyshev Polynomials*, Hoboken, N.J.: Wiley, 1974.

Approximation of the Density Function for the Latest Critical Time

Now we consider $f_{X_i}^{LCT}$. Each iteration of the latest critical time calculation involves the convolution of f_{X_i} with $f_{X_i}^{MOS}$. The convolution of f_{X_i} with $f_{X_i}^{MOS}$ is performed using the algorithm of Nicholas Hale and Alex Townsend.[61] This algorithm ensures that the calculations are performed efficiently in the transformed space.

Approximation of the Density Function for the Critical Predecessor

The double integral for $P_c(v_i, v_j)$ is performed using a two-dimensional Chebyshev representation,

$$P_C(v_i, v_j) = \sum_{\alpha=1}^{M} \sum_{\beta=1}^{M} a_{\alpha\beta}\, T_\alpha(t_i)\, T_\beta(t_j), \tag{5.18}$$

where the coefficients $a_{\alpha\beta}$ are computed as shown below:

$$a_{\alpha\beta} = \frac{\int_{-1}^{1} \int_{-1}^{1} w(t_i, t_j)\, P_c(v_i, v_j)\, T_\alpha(t_i)\, T_\beta(t_j)\, dt_i\, dt_j}{\int_{-1}^{1} \int_{-1}^{1} w(t_i, t_j)\, T_\alpha(t_i)\, T_\beta(t_j)\, dt_i\, dt_j}, \quad \text{for } \alpha, \beta \in \{1, \ldots, M\}, \tag{5.19}$$

where the weight $w(t_i, t_j)$ is given by

$$w(t_i, t_j) = \frac{1}{\sqrt{1 - t_i^2}\sqrt{1 - t_j^2}}. \tag{5.20}$$

Once the density functions for the critical predecessors of each node are computed, this information is combined to calculate the criticality of each node, i.e. $C_{\text{Node}}(v_i)$ for each $v_i \in G$, using Equation 5.6. This completes the construction of the Chebyshev spectral method for computing the criticality index. With the inclusion of each successive basis function in the Chebyshev approximation, the error for the approximation decreases as $O(e^{-M})$, where M is the number of terms in the Chebyshev series. This compares favorably with the second-order method with trapezoidal integration (which converges as $O(M^{-2})$) and the Monte Carlo method (which converges as $O(1/\sqrt{M})$).

Both the trapezoidal method and the spectral method compute much faster than simulation methods, opening the critical path analysis to very large graphs. We have implemented both in

---

[61] Nicholas Hale and Alex Townsend, "An Algorithm for the Convolution of Legendre Series," *SIAM Journal on Scientific Computing*, Vol. 36, No. 3, 2014.

code. The trapezoidal method works well in Python, and our Python code incorporates it. But because the tools in Python do not readily support some of the calculations needed for the manipulation of Chebyshev polynomials (as of 2020), we have implemented the spectral method in code written in Matlab.

6. Conclusions

As part of the digital engineering initiative, acquisition programs in the U.S. Department of Defense are increasingly holding authoritative, up-to-date information on all aspects of programs in digital form. These data provide rich opportunities for more-sophisticated and more-rigorous analysis, of both engineering activities and program management activities. The work presented in this report summarizes algorithms delivered to the GBSD program office for better managing program execution activities. The algorithms help project managers consider "what if" scenarios and visualize potential pitfalls as a project progresses. Project managers can also use these algorithms to perform sensitivity analyses on individual activities to understand which activities have the greatest potential to derail a project. Although they were developed for UCS activities in particular, the concepts and algorithms are applicable to any set of interdependent activities in a complex program.

The work builds on the pioneering work of PERT/CPM. These methods, which were developed in the late 1950s, have been continuously refined in the years since. In this report, we drew on this existing work for many of the methods but further built on this literature in two significant ways. The first was to examine the dependence of critical paths on PERT graph topology, and the second was to develop novel, very fast numerical techniques that permit critical path analysis on very large PERT graphs.

Novel Topological Insights

The critical path and degree of slack depend on graph topology. When managers have the opportunity to change the structure of activities, they can better manage where schedule risk lies in program execution. We examined three classes of graph topology:

- a simple topology defined by the number of parallel chains in the graph
- random graphs (of Erdős-Rényi topology)
- scale-free graphs.

The simple chain topology is elementary but contains structural elements common to many PERT graphs. It is intuitive and contains the basic structural forms that a manager can readily adjust. The random and scale-free graph cases are common topologies in naturally occurring graphs. Many large graphs present one or the other of these two topologies; therefore, they are useful baselines for analysis.

Simple Chain Topology

For the simple chain topology, our analysis shows that although parallelizing development chains can theoretically reduce the overall time, it does not do so as much as one would expect,

47

because the likelihood of having at least one bad chain becomes higher the more chains there are. There is a trade-off: Parallelizing development reduces the variance more than expected but also increases the end-to-end time more than expected. Thus, although working tasks in parallel remains desirable, the project manager's expectations for time savings should be tempered. This issue is compounded when a project has multiple sets of subprocesses proceeding in parallel.

Random Graphs of Erdős-Rényi Topology

For Erdős-Rényi topology (random graphs), our analysis shows the same trade-off between variance and average completion time as seen in the simple chain topology, though the effect is more subtle. In this case, the presence of interlinking paths makes it less likely that a single chain will dominate, because the chains are no longer completely independent.

Scale-Free Graph Topology

For the scale-free graph topology, our analysis shows that graphs with a single, dominant longest path are much more likely to complete on time than graphs with several parallel paths that may each be the critical path. This result means that, for scale-free graphs, the project manager can reduce the overall completion time with only a marginal increase in the corresponding risk by placing important tasks in series with one another. We caution that this result may not be robust to different assumptions: Each of the nodes considered here had identical beta distribution parameters, whereas a real graph is more likely to have a mix of different parameters that may be correlated with degree distribution.

Novel Numerical Methods

Although Monte Carlo techniques are useful for finding the critical path in a small PERT graph, as graph size increases, these methods are too computationally slow to be of practical use. To overcome this limitation, we presented a novel numerical approach that is many orders of magnitude faster. We developed two methods. Both are semi-analytic rather than simulations. The fastest, and preferred, method transforms the problem into a new space with Chebyshev polynomials as a basis. The relevant calculations are performed in that space; then, the results are inverted back to the original space. The other method uses trapezoidal integration but is preferred only when software packages limit the ability to manipulate Chebyshev polynomials.

Resource Constraints and Future Analysis

Program managers are conscious not only of the time requirements but also of the resources necessary to complete their projects. Resource limitations can severely delay a project if they are not managed effectively. We described how slack in a task network can be manipulated to reduce project completion time; to reduce the overall need for resources; or to respond to other needs, such as start-up constraints, changes in workforce availability or capacity, or cash flows. However, current methods rely on either expected-value algorithms, which may not sufficiently capture uncertainty, or Monte Carlo simulations, which require excessive amounts of runtime for large, complex projects. Future work in this area could extend the novel numerical methods described in this report to closely approximate solutions that could prove invaluable to the program manager.

The algorithms that have been delivered to the GBSD program office and this explanation of those approaches should aid in the management of complex programs, especially those following digital engineering practices.

Appendix A. Equivalence of Representations

PERT/CPM graphs originally represented tasks or activities on arcs (AoA) (links) rather than activities on nodes (AoN). In the AoA representation, the nodes represent task completion milestones, task initiation, or both. For example, in Figure A.1, node 1 represents the initiation of a project and node 2 the completion of task A. The task occurs between nodes 1 and 2 and can be written as either activity A or activity (1, 2). Since task B cannot begin until activity A is complete, the start node for B, node 2, is identical to the completion node for task A. Finally, Task B is completed at node 3.[62]

Figure A.1. A Simple AoA Example

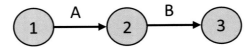

For a given project, it is possible to create AoA and AoN representations that are equivalent in the sense that they represent the same activities and the same precedence relationships. Most importantly, they yield exactly the same results. Why, then, have recent PERT/CPM project abstractions, including the one we present, used the AoN approach, as described in Chapter 2?

To explain this, let us take an example network from Chapter 3. Table A.1 shows the task, predecessors, and successors for this PERT network. Figure A.2 provides the AoN diagram of the network. Creating the AoA version has two small twists. They revolve around some activities that have multiple predecessors.

First, consider activity H. It must be preceded by two activities, F and G. In the AoA approach, there will be a node for the end of F (number 6 in Figure A.3) and a node for the end of G (number 7 in Figure A.3). The problem is that only one of these can double as the node that starts H. Let the node representing the end of G also be the node that is the start of H. To preserve the precedence relationship, it is necessary to add a dummy link between the end of F (node 6) and the combined end of G/start of H (node 7), which requires zero time to complete. Resolving these multiple predecessor relationships complicates the formulation of the AoA graph.

---

[62] Of course, node 2 could be split into two nodes, one for the completion of A and the other for the start of B. The link between these would have a dummy link with zero time. However, we desire to create an AoA graph with as few nodes and links as possible in order to reduce the size of the graph and the associated computation time. This also shows that there are multiple ways of representing an AoA graph while there is only one way to represent the AoN equivalent.

Table A.1. Data for an Example Project

Task	Predecessors	Successors
A	None	B, C, and D
B	A	D
C	A	F and G
D	A and B	D
E	D	I
F	C	H
G	D	H
H	F and G	I
I	E and H	None

Figure A.2. The AoN Graph for an Example Project

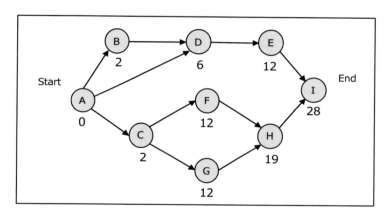

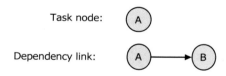

Task node: A

Dependency link: A → B

Figure A.4 shows a version of the AoA graph for this example project. It is one of many, since, for example, we could have had H start at the end of F and a dummy link from node 7 to node 6. Note also the odd extra dummy node that starts at the end of A (node 2) and ends at the beginning of D (node 3) to capture the multiple precedence of tasks A and B. It is clear that this dummy node could just as well be dropped.

Figure A.3. Resolving Multiple Predecessors in an AoA Graph

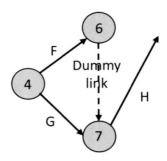

Figure A.4. An AoA Graph for the Example Project

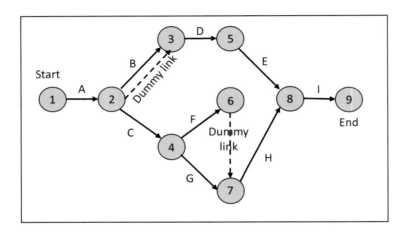

Milestone: ①

Task: ① —A→ ②

Dummy link: ① - - - → ②

In addition to these advantages of the AoN approach (i.e., it is easier to formulate, produces a unique graph, and does not require extra dummy nodes or links), we have chosen the AoN in this report because the representation more closely matches the MBSE data construct.

Appendix B. The PERT and CPM Algorithms

The PERT and CPM algorithms were developed to determine key details needed by project managers to effectively administer large-scale projects, including the following:[63]

- the duration of the project if average task durations are achieved
- the latest start times of each task to meet the project's expected completion time
- the earliest task start times that satisfy task precedence requirements
- the tasks that must be completed on time to avoid delaying project completion; these are the tasks on the *critical path*
- for those tasks not on the critical path, the amount of time that they could be delayed without negatively affecting the project completion time; this is often called the *task slack time*
- the probability that the project will be completed at a specific time
- the cost to complete the project by a specific time
- the trade-off between cost and project completion time
- monitoring the project cost against the budget as the project moves forward.

To determine this information, the original PERT and CPM algorithms used an expected-value approach. While the expected value only approximates the above metrics, a more exact approach using a stochastic method is considerably more complicated, as shown in Chapters 3 to 5 of this report. In this appendix, we review the expected-value PERT and CPM algorithms as a foundation for the more advanced stochastic algorithms.

The PERT Expected-Value Algorithm

The PERT algorithm originally developed by the U.S. Navy relies on estimates of subject-matter experts to characterize task completion times. An average time to completion can be given, or, to include variation in the calculation, subject-matter experts can provide optimistic (minimum), most likely, and pessimistic (maximum) times for each task. With these values, a simple statistical formula is used to determine the average time to complete the task, as shown in Table B.1.[64]

---

[63] The material in this appendix may also be found in project management or operations research texts and articles, such as Erik Leuven Demeulemeester and Willy S. Herroelen, *Project Scheduling: A Research Handbook*, Boston, Mass.: Springer, 2002; Adedeji B. Badiru, *Project Management: Systems, Principles, and Applications*, 2nd ed., Boca Raton, Fla.: CRC Press, 2019; and Hillier and Lieberman, 2014.

[64] A beta distribution is used to fit the optimistic, most likely, and pessimistic values and calculate the mean and the variance. For this distribution, the approximate value of the mean is (optimistic + 4 × most likely + pessimistic)/6 and for the variance is ((pessimistic – optimistic)/6)^2. The mathematics for this can be found in Rafael Herrerías Pleguezuelo, José García Pérez, and Salvador Cruz Rambaud, "A Note on the Reasonableness of PERT Hypotheses," *Operations Research Letters*, Vol. 31, No. 1, 2003.

Table B.1. Task Durations

Task	Time (months)				
	Optimistic	Most Likely	Pessimistic	Average	Variance
A	1	2.00	3	2	0.11
B	2	3.50	8	4	1.00
C	6	9.00	18	10	4.00
D	4	5.50	10	6	1.00
E	1	4.50	5	4	0.44
F	4	4.00	10	5	1.00
G	4	6.75	11	7	1.36
H	5	8.00	17	9	4.00
I	3	7.50	9	7	1.00

Once the average task durations are determined, the algorithm traverses the tasks in precedence order, calculating the earliest start time for each task. Figure B.1 shows these earliest start times under each of the task nodes in the network. For example, starting with the first task, A, the earliest start time will be at the very beginning of the project, or month 0. Next, nodes B and C are only reliant on the completion of task A.[65] To get their earliest start time, we take the earliest start time for A (0) and add it to the average duration of task A (2). Thus, the earliest start times for B and C are at time 2 months. As another example from Figure B.1, consider task D. Task D is dependent on the completion of both task A and task B. Each will finish at a time calculated as the sum of their earliest start time and their average duration. Thus, as we already have seen, task A completes at 2 months, but task B completes at 6 months. This means that the earliest start time for D is the maximum end time of all of its preceding tasks, or 6 months after the start of the project. A similar process is used to calculate the earliest start time of all tasks with multiple predecessors. The final average earliest end time for the project is 35 months, adding task I's duration of 7 months to its earliest start time of 28 months.

---

[65] Task D is also dependent on task A, but the additional dependence on task B requires that the earliest start time for task D be calculated after the earliest start time for task B has been calculated.

Figure B.1. Notional Task Dependency Graph with Average Earliest Start Times

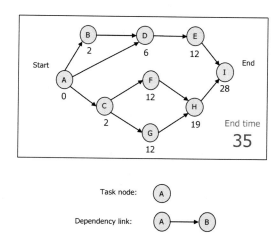

The next step in the PERT algorithm is to determine the critical path and the amount of slack for the tasks that are not on the critical path. This calculation is performed by determining the latest start time for each of the tasks that will not delay project completion. The rationale is that any task that has the same earliest and latest start times will delay the whole project if it is delayed and so must be on the critical path. To find the latest start times, we work backward from the project end time using a similar approach as above. For example, the latest start time for task I is 35 (the project end time) less the duration of project I (7), yielding 28. The algorithm continues backward using the task average durations and the latest start times. Figure B.2 shows both the earliest start and latest start times parenthesized beneath each task node in the graph.

Figure B.2. Notional Task Dependency Graph with Average Earliest Start Times, Average Earliest Completion Times, and the Critical Path

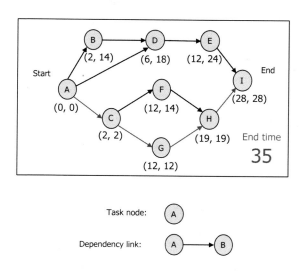

Figure B.2 also depicts the critical path for the project by the links colored blue. Following the tasks for which the earliest and latest start times are equal, the critical path goes through the following tasks: A, C, G, H, and I. The tasks not listed on the critical path have slack times as listed in Table B.2. Tasks with some slack offer a way for project managers to possibly reduce the project end time. Resources assigned to tasks with slack may be able to be reallocated to tasks on the critical path, thereby reducing their completion duration as well as the project completion time.

Table B.2. Task Durations, Critical Path, and Slack

Task	Time (months)			Average	Variance	Critical Path	Slack
	Optimistic	Most Likely	Pessimistic				
A	1	2.00	3	2	0.11	Yes	0
B	2	3.50	8	4	1.00	No	12
C	6	9.00	18	10	4.00	Yes	0
D	4	5.50	10	6	1.00	No	12
E	1	4.50	5	4	0.44	No	12
F	4	4.00	10	5	1.00	No	2
G	4	6.75	11	7	1.36	Yes	0
H	5	8.00	17	9	4.00	Yes	0
I	3	7.50	9	7	1.00	Yes	0

Note that when the project is actually executed, the project duration may be determined by a set of tasks that were not on the planned critical path. For example, if the optimistic time is achieved for tasks A, C, G, H, and I (the tasks on the planned critical path), that path's completion time will be only 20 months, while, if the pessimistic time is required for tasks B, D, and E, the path that also includes A and I will require a time of 27 months. What, then, is the likelihood of some task being part of the longest path when the project is executed? To answer this question requires the more-sophisticated methods described in Chapters 3 and 5.

Approximating the Probability of Meeting a Target Project End Time

An approximation using the PERT algorithm can be made for the probability that the project completion will be less than a specific time. For example, while the expected conclusion is 35 months, suppose we would like to know what the probability is that the actual time is less than 38 months. To make this kind of estimation, three simplifying assumptions are needed:

- First, we assume that the tasks on the critical path will indeed be the tasks on the longest path. This is certainly a limiting factor, since, as we have noted, other paths could be longer during the execution of the project.

- Second, the durations of the tasks on the critical path are statistically independent. Although this is likely, there may be circumstances in which conditions that cause a delay in one task on the critical path may also affect another task on that path.
- Finally, we assume that variation in the project completion time is normally (Gaussian) distributed. This estimation is dependent on the number of tasks in the critical path. The greater the number, the more closely the normal distribution will approximate the actual distribution.

Using these assumptions, we can calculate the probability that the end time is less than 38 months. For this, we need the mean for the critical path (35 months) and the variances for the completion times for tasks on the critical path (given above). Summing these variances yields 10.47 for a standard deviation of 3.24. Applying the normal distribution gives the approximate probability that the project will take less than 38 months as 82.3 percent. This method can be used to provide the probability that the project will end for a full range of times, as shown in Figure B.3.

Figure B.3. Probability of Project End Time

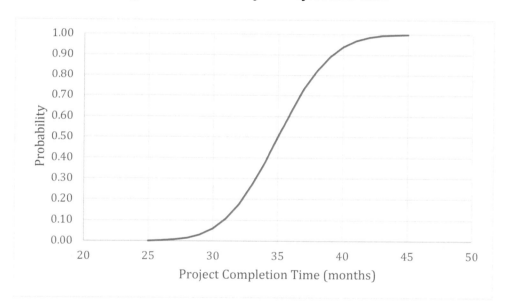

Cost-Time Trade-Offs

Project managers are frequently concerned about reducing the time needed to complete a project and the associated required resources. This leads to the consideration of time-cost trade-offs among the project tasks. For many tasks, there may be an ability to speed up the completion by supplying additional people, having overtime hours, or setting up multiple shifts. Reductions might also be achieved by employing more highly qualified workers or providing a greater quantity of supplies. Each of these means comes at a premium cost. If the advantage of completing the project earlier, such as a monetary bonus, is sufficiently large or the disadvantage of late completion, such as a financial penalty, is large, completing the task early at a high cost may be warranted.

For example, suppose the bonus for completing the project above in less than 32 months is $2,000,000. Is it cost-effective for the project manager to pay the premium to reduce the time? To answer this question, consider the time and costs in Table B.3. Tasks, as well as average and minimum (optimistic) task completion times, are as shown in Table B.2. The cost for completing each task in the average amount of time is given, along with the premium cost for each month of reduction up to and including the minimum amount of time for the task.

Table B.3. Project Completion Times and Costs

Task	Time (months)		Cost (thousands of dollars)	
	Average	Minimum	Average Time	Each Month Reduction
A	2	1	1,850	1,387.50
B	4	2	1,050	393.75
C	10	6	5,560	834.00
D	6	4	2,800	700.00
E	4	1	910	341.25
F	5	4	1,200	360.00
G	7	4	2,930	627.86
H	9	5	4,550	758.33
I	7	3	3,250	696.43

The cost-time trade-off in this example is linear for each possible month that can be saved. In general, costs to reduce task time will not be linear, since a decrease of one month will likely cost less than the cost for the second month's reduction. Subsequent decreases in time will likely cost even more. Linearity of the marginal time reduction allows us to solve this problem with one of two techniques.

The first technique uses a simple marginal cost. At each step, the critical path is determined. From the critical path, the task with the least monthly reduction cost is selected. For example, the initial critical path ACGHI has G with the smallest cost reduction. Therefore, for a cost of $627,860, the project time can be reduced from 35 to 34 months. Once a task is reduced by one month, it is possible that a new critical path will be encountered—thus the need to calculate the critical path at each step. It is also necessary to determine whether the one-month decrease in a task creates multiple critical paths, possibly requiring reductions from more than one task. Reductions could continue at ever-increasing costs until a targeted time for the project is reached or the critical path has each of its tasks at the minimum time possible. In this example, the cost for reaching the targeted time is just more than $2,243,000. When comparing this cost with the $2,000,000 bonus, the project manager decides that it is not economically desirable to reduce the project time. While this algorithm is effective for small projects, recalculating the critical path(s) at each step of a large, complex project becomes unwieldy.

A second approach is to formulate and execute a linear program with a variable for each task representing the amount of time saved. The objective function, then, sums the product of the reduction cost and the time reduction variable for each task. Network flow constraints are added to ensure that the precedence relationships are maintained and to calculate the project's final duration. Capturing the project duration as a variable allows a constraint to be added, limiting the project to a specified target time. While it is typically faster than the marginal approach for large-scale projects, the solution time for the linear program can be lengthy. Furthermore, the linearity of the time-cost relationship is likely to be unrealistic. Convex functions and associated optimizers can be used instead, having potentially little impact on the solution time while more accurately portraying the time-cost trade-off.

Determining Possible Budget Needs by Time Period

The project manager may desire to know what flexibility exists for the project cost flow over the duration of the project. For this, we use the earliest and latest task start times based on the average task duration shown in Table B.3. We also assume that the cost for a task using the average time can be spread evenly over that duration. Given the slack in the schedule, the project need not start paying for a task at its earliest start time but must pay the cost by the latest start time. This then creates a range of possible budget and payment schedules for the project.

Figure B.4 displays this range graphically. Each month of the project, from inception to month 35, is shown on the x-axis. The cumulative project cost is on the y-axis. The blue line provides the monthly cost if each task is begun at its earliest start time. The red line is for delaying any task with slack until its latest start time. While the final cost is the same either way, the gap between the blue line and the red line shows possible resource flows that could be used by the project manager.

Figure B.4. Budget and Payment Flexibility

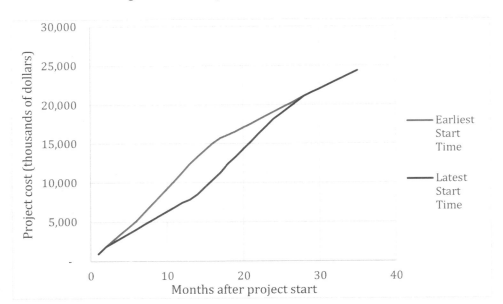

Summary

This brief description of the basic PERT and CPM algorithms and some of their extensions demonstrates how these simple tools can provide valuable information for the project manager and other decisionmakers. In our discussion, we have described how these algorithms can be used during the planning process of a project, but they can also be used during project execution to monitor the success of the plan. For example, after some tasks have been completed, when some are in progress and others have not been started, these algorithms can be processed again to determine new estimates of project completion time, the critical path, or additional expenditures that may be needed to keep the project on schedule. Furthermore, changes to task durations or cost may mean that both the project schedule and the resource flow algorithms need to be rebalanced through multiple runs of the algorithms.

As a final note, we remind the reader that these algorithms are simplifications of actual project management challenges. Expected values and linearity were assumed in these original algorithms, which may produce inaccurate, biased results. The stochastic algorithms presented in the body of this report provide more-exact results.

Appendix C. Proof That Convolution of Independent, Identical Beta Distributions Converges to a Normal Distribution

In this appendix, we prove that the convolution of a sequence of independent and identically distributed (iid) beta distributions converges to a normal distribution. Without loss of generality, we prove the result for the beta distribution defined with compact support on $[0,1]$.

Recall that the PDF of the beta distribution (with compact support on the interval $[0,1]$) is given by

$$f_{X_B}(x; \alpha, \beta) = \frac{1}{B(\alpha, \beta)} x^{\alpha-1} (1-x)^{\beta-1}, \tag{C.1}$$

where $x \in [0,1]$, and $B(\alpha, \beta)$ is expressed in terms of the gamma function,[66] as

$$B(\alpha, \beta) = \frac{\Gamma(\alpha + \beta)}{\Gamma(\alpha)\, \Gamma(\beta)}. \tag{C.2}$$

The characteristic function of a random variable $X: \Omega \to [0,1]$, denoted by $\varphi_X: \mathbb{R} \to \mathbb{C}$, is defined by

$$\varphi_X(t) = E[e^{itX}] = \int_{\mathbb{R}} e^{itx}\, 1_{\{x \in \Omega\}}\, d\, F_X(x) = \int_{\Omega} e^{itx}\, f_X(x)\, dx. \tag{C.3}$$

Note that the characteristic function is the Fourier transform of X. The characteristic function of the beta distribution has a convenient representation in terms of the Kummer confluent hypergeometric function (KHF). The KHF is defined by

$$M(a, b, z) = \frac{\Gamma(b)}{\Gamma(a)\, \Gamma(b-a)} \int_0^1 e^{zu}\, u^{a-1}\, (1-u)^{b-a-1}\, du, \tag{C.4}$$

where $0 < Re(a) < Re(b)$, and $z \in \mathbb{C}$. Using the KHF, the characteristic function of the beta distribution can be written as follows:

---

[66] Consider the integral $\Gamma(z) = \int_0^\infty x^{z-1} e^{-x}\, dx$ defined for $Re(z) > 0$. The *gamma function* is defined as the analytic continuation of this integral to the entire complex plane, except for the negative integers.

$$\varphi_{X_B}(t) = \int_{\mathbb{R}} e^{itx} \, 1_{\{x \in [0,1]\}} \, d \, F_{X_B}(x; \alpha, \beta) = \int_0^1 e^{itx} \, f_{X_B}(x; \alpha, \beta) \, dx = M(a, a+b, it). \quad \text{(C.5)}$$

Using the power series expansion of KHF, the characteristic function is written as

$$\varphi_{X_B}(t) = M(a, a+b, it) = \sum_{k=0}^{\infty} \frac{\alpha^{(k)} \, (it)^k}{(\alpha + \beta)^{(k)} \, k!} = 1 + \sum_{k=1}^{\infty} \left(\prod_{j=0}^{k-1} \frac{\alpha + j}{\alpha + \beta + j} \right) \frac{(it)^k}{k!}, \quad \text{(C.6)}$$

where $\alpha^{(k)} = \prod_{j=0}^{k-1} (\alpha + j)$. This result will be used in the subsequent discussion to explicitly prove that the sum of iid beta distributions converges to a normal distribution.

We will now cite an important theorem by P. Lévy that relates a sequence of probability distributions with the corresponding characteristic functions.

Theorem (Lévy's Continuity Theorem):[67] Let $\{X_k\}_k$ be a sequence of random variables that converges in distribution to a random variable X,[68] and let $\{\varphi_{X_k}\}_k$ be the corresponding characteristic functions. Then, $X_k \to X$ in distribution if and only if $\varphi_{X_k}(t) \to \varphi_X(t)$ pointwise, i.e., for all $t \in \mathbb{R}$.

We will now prove the main result of this appendix.

Proposition: Let $\{X_{B_k}\}_{k=1}^{N}$ be a sequence of iid random variables that are beta-distributed with mean μ and variance σ^2, and let $\bar{X}_{B_N}$ denote the average $\left(\sum_{k=1}^{N} X_{B_k} \right) / N$. Then, $\bar{X}_{B_N}$ converges (in distribution) to $N(\mu, \sigma^2 / N)$, i.e., the normal distribution with mean μ and variance σ^2.

Proof: Let $\{X_{B_k}\}_{k=1}^{N}$ be a sequence of independent random variables that are beta-distributed for fixed α and β. Hence, each random variable has mean μ and variance σ^2, given by

$$\mu = \frac{\alpha}{\alpha + \beta}, \qquad \sigma^2 = \frac{\alpha \beta}{(\alpha + \beta)^2 \, (\alpha + \beta + 1)}. \quad \text{(C.7)}$$

Now, we define the random variable S_{B_N} as

$$S_{B_N} = \frac{\sum_{k=1}^{N} X_{B_k} - N \mu}{\sqrt{N} \, \sigma} = \frac{1}{\sqrt{N}} \sum_{k=1}^{N} \frac{X_{B_k} - \mu}{\sigma}. \quad \text{(C.8)}$$

---

[67] James Norris, "Advanced Probability," lecture notes, November 15, 2019.

[68] Let $\{X_k\}_{k=1}^{N}$ be a sequence of real-valued random variables with corresponding CDFs denoted by F_N. Recall that $\{X_k\}_k$ converges in distribution to a random variable X if $\lim_{N \to \infty} F_N(x) = F(x)$, for all $x \in \mathbb{R}$ where F is continuous, where F denotes the CDF of the random variable X.

Set $Y_{B_k} = \frac{X_{B_k} - \mu}{\sigma}$ so that $E[Y_{B_k}] = 0$ and $\text{Var}[Y_{B_k}] = 1$ for all $k \in \{1, \dots, N\}$. Hence, the characteristic function of S_{B_N} can be written as

$$\varphi_{S_{B_N}}(t) = E\left[e^{it\frac{1}{\sqrt{N}}\sum_{k=1}^{N} Y_{B_k}}\right] = E\left[\prod_{k=1}^{N} e^{it\frac{1}{\sqrt{N}} Y_{B_k}}\right] = \prod_{k=1}^{N} E\left[e^{it\frac{1}{\sqrt{N}} Y_{B_k}}\right] = \prod_{k=1}^{N} \varphi_{Y_{B_k}}\left(\frac{t}{\sqrt{N}}\right), \tag{C.9}$$

where the third equality holds because of the independence of the Y_{B_k}. Since the Y_{B_k} are identically distributed, we can write $\varphi_{S_{B_N}}(t)$ as

$$\varphi_{S_{B_N}}(t) = \prod_{k=1}^{N} \varphi_{B_k}\left(\frac{t}{\sqrt{N}}\right) = \left[\varphi_{Y_{B_k}}\left(\frac{t}{\sqrt{N}}\right)\right]^N. \tag{C.10}$$

Now we note that the characteristic function for each of the Y_{B_k} can be expressed in terms of the KHF power series as

$$\varphi_{Y_{B_k}}(t) = 1 + \sum_{n=1}^{\infty} \left(\prod_{j=0}^{n-1} \frac{\alpha + j}{\alpha + \beta + j}\right) \frac{(it)^n}{n!}. \tag{C.11}$$

Using this representation, we may write $\varphi_{Y_{B_k}}\left(\frac{t}{\sqrt{N}}\right)$ to second order as

$$\varphi_{Y_{B_k}}\left(\frac{t}{\sqrt{N}}\right) \sim 1 + \left[\frac{\alpha/(\alpha+\beta) - \mu}{\sqrt{N}\,\sigma}\right] it - \left[\frac{1}{N\,\sigma^2}\left(\frac{\alpha}{\alpha+\beta}\right)\left(\frac{\alpha+1}{\alpha+\beta+1}\right) - \frac{\mu^2}{N\,\sigma^2}\right] t^2 + \text{h.o.t.}, \tag{C.12}$$

where h.o.t. denotes higher-order terms, which are $o\left(\frac{t^2}{N}\right)$. The first-order term in the expression above vanishes, since $\mu = \frac{\alpha}{\alpha+\beta}$ for the beta distribution, and the coefficient of the second term simplifies to $\frac{1}{2N}$. Hence,

$$\varphi_{Y_{B_k}}\left(\frac{t}{\sqrt{N}}\right) \sim \left(1 - \frac{t^2}{2N} + \text{h.o.t.}\right), \tag{C.13}$$

and, therefore, as $N \to \infty$:

$$\lim_{N\to\infty} \varphi_{S_{B_N}}(t) = \lim_{N\to\infty}\left(1 - \frac{t^2}{2N} + \text{h.o.t.}\right)^N = e^{-\frac{t^2}{2}}, \qquad \forall t \in [0,1] \text{ (i.e., pointwise)}. \tag{C.14}$$

The right-hand side of the expression above is the characteristic function of $N(0,1)$, i.e., the standard normal distribution. Therefore, by the Lévy continuity theorem, the sequence $\{S_{B_N}\}$ converges to $N(0,1)$ as $N \to \infty$. For sufficiently large N, this implies that

$$\sum_{k=1}^{N} X_{B_k} \sim N(N\mu, N\sigma^2). \qquad (\text{C.15})$$

Finally, it follows that

$$\frac{\sum_{k=1}^{N} X_{B_k}}{N} \to N\left(\mu, \frac{\sigma^2}{N}\right), \quad \text{as } N \to \infty. \qquad (\text{C.16})$$

Appendix D. Derivation of Maximum Order Statistic Formula

Equation 5.4 in Chapter 5 finds the PDF of MOS given a set of individual PDFs—the probability distribution of the maximum of a set of random variates. We compute it by noting that the probability that any given random variate V_i is less than x is the integral from negative infinity to x of the probability distribution:

$$P(V_i < x) = \int_{-\infty}^{x} PDF_i(x')dx'. \tag{D.1}$$

Then, the probability that *all* of the random variates are less than x is just the product of each probability:

$$P(V_i < x \,\forall\, i) = \prod_i \int_{-\infty}^{x} PDF_i(x')dx'. \tag{D.2}$$

Put another way, Equation D.2 is the CDF for the MOS. Finding the PDF given the CDF is then a matter of taking the derivative with respect to x:

$$PDF_{max}(x) = \frac{d}{dx} \prod_i \int_{-\infty}^{x} PDF_i(x')dx'. \tag{D.3}$$

Replacing $x \rightarrow t$ and changing the lower limit of integration from $-\infty$ to 0 yields Equation 5.4.

References

Air Force Instruction 63-125, *Nuclear Certification Program*, Washington, D.C.: Headquarters, Department of the Air Force, January 16, 2020.

Badiru, Adedeji B., *Project Management: Systems, Principles, and Applications*, 2nd ed., Boca Raton, Fla.: CRC Press, 2019.

Bollobás, Béla, Christian Borgs, Jennifer Chayes, and Oliver Riordan, "Directed Scale-Free Graphs," *Proceedings of the Fourteenth Annual ACM-SIAM Symposium on Discrete Algorithms*, 2003, pp. 132–139.

———, "Percolation on Dense Graph Sequences," *Annals of Probability*, Vol. 38, No. 1, January 2010, pp. 150–183.

Boyd, John P., *Chebyshev and Fourier Spectral Methods*, 2nd ed., Mineola, New York: Dover Publications, Inc., 2000.

Demeulemeester, Erik Leuven, and Willy S. Herroelen, *Project Scheduling: A Research Handbook*, Boston, Mass.: Springer, 2002.

Department of Defense Instruction 8510.01, *Risk Management Framework (RMF) for DoD Information Technology (IT)*, Washington, D.C.: U.S. Department of Defense, March 12, 2014, incorporating change 2, July 28, 2017.

Diestel, Reinhard, *Graph Theory*, 5th ed., Berlin, Germany: Springer, 2017.

Dodin, Bajis M., "Approximating the Distribution Functions in Stochastic Networks," *Computers & Operations Research*, Vol. 12, No. 3, 1985, pp. 251–264.

Dodin, Bajis M., and Salah E. Elmaghraby, "Approximating the Criticality Indices of the Activities in PERT Networks," *Management Science*, Vol. 31, No. 2, February 1985, pp. 207–223.

Easley, David, and Jon Kleinberg, *Networks, Crowds, and Markets: Reasoning About a Highly Connected World*, New York: Cambridge University Press, 2010.

Engwall, Mats, "No Project Is an Island: Linking Projects to History and Context," *Research Policy*, Vol. 32, No. 5, May 2003, pp. 789–808.

———, "PERT: Polaris and the Realities of Project Execution," *International Journal of Managing Projects in Business*, Vol. 5, No. 4, 2012, pp. 595–616.

———, *PERT, Polaris, and the Realities of Project Execution*, Stockholm, Sweden: KTH Royal Institute of Technology, 2015.

Ghomi, S. M. T. Fatemi, and E. Teimouri, "Path Critical Index and Activity Critical Index in PERT Networks," *European Journal of Operational Research*, Vol. 141, No. 1, August 2002, pp. 147–152.

Hale, Nicholas, and Alex Townsend, "An Algorithm for the Convolution of Legendre Series," *SIAM Journal on Scientific Computing*, Vol. 36, No. 3, 2014, pp. A1207–A1220.

Hillier, Frederick S., and Gerald J. Lieberman, *Introduction to Operations Research*, 10th ed., New York: McGraw-Hill Education, 2014.

Hörmander, Lars, *The Analysis of Linear Partial Differential Operators I: Distribution Theory and Fourier Analysis*, Berlin: Springer, 2015.

Jungnickel, Dieter, *Graphs, Networks, and Algorithms*, 4th ed., Berlin: Springer-Verlag, 2013.

Kahn, Arthur B., "Topological Sorting of Large Networks," *Communications of the Association for Computing Machinery*, Vol. 5, No. 11, 1962, pp. 558–562.

Kelley, James E., Jr., and Morgan R. Walker, "The Origins of CPM: A Personal History," *PM Network*, Vol. 3, No. 2, February 1989, pp. 7–22.

MacCrimmon, K. R., and C. A. Ryavec, *An Analytical Study of the PERT Assumptions*, Santa Monica, Calif.: RAND Corporation, RM-3408-PR, 1962. As of July 9, 2021: https://www.rand.org/pubs/research_memoranda/RM3408.html

Martin, J. J., "Distribution of the Time Through a Directed, Acyclic Network," *Operations Research*, Vol. 13, No. 1, 1965, pp. 46–66.

Moder, Joseph J., and Cecil R. Phillips, *Project Management with CPM and PERT*, New York: Van Nostrand-Reinhold, 1970.

Norris, James, "Advanced Probability," lecture notes, November 15, 2019.

Office of the Deputy Assistant Secretary of Defense for Systems Engineering, *Department of Defense Digital Engineering Strategy*, Washington, D.C.: U.S. Department of Defense, June 2018.

Palacio, Juan D., and Olga L. Larrea, "A Lexicographic Approach to the Robust Resource-Constrained Project Scheduling Problem," *International Transactions in Operational Research*, Vol. 24, No. 1/2, January–March 2017, pp. 143–157.

Pleguezuelo, Rafael Herrerías, José García Pérez, and Salvador Cruz Rambaud, "A Note on the Reasonableness of PERT Hypotheses," *Operations Research Letters*, Vol. 31, No. 1, 2003, pp. 60–62.

Rivlin, Theodore J., *The Chebyshev Polynomials*, Hoboken, N.J.: Wiley, 1974.

Roper, Will, *There Is No Spoon: The New Digital Acquisition Reality*, September 18, 2020.

Rudin, Walter, *Real and Complex Analysis*, 3rd ed., New York: McGraw-Hill, 1987.

Sapolsky, Harvey M., *The Polaris System Development: Bureaucratic and Programmatic Success in Government*, Cambridge, Mass.: Harvard University Press, 1972.

Stein, Elias M., *Harmonic Analysis: Real-Variable Methods, Orthogonality, and Oscillatory Integrals*, Princeton, N.J.: Princeton University Press, 1993.

Targiel, Krzysztof S., Maciej Nowak, and Tadeusz Trzaskalik, "Scheduling Non-Critical Activities Using Multicriteria Approach," *Central European Journal of Operations Research*, Vol. 26, No. 3, September 2018, pp. 585–598.

Tarjan, Robert E., "Edge-Disjoint Spanning Trees and Depth-First Search," *Acta Informatica*, Vol. 6, No. 2, 1976, pp. 171–185.

Trefethen, Lloyd N., "Approximation Theory and Approximation Practice," *SIAM*, Vol. 164, 2019.

Van Slyke, Richard M., "Monte Carlo Methods and the PERT Problem," *Operations Research*, Vol. 11, No. 5, September–October 1963, pp. 839–860.

Zimmerman, Phil, Tracee Gilbert, and Frank Salvatore, "Digital Engineering Transformation Across the Department of Defense," *Journal of Defense Modeling and Simulation: Applications, Methodology, Technology*, Vol. 16, No. 4, 2019, pp. 325–338.